From Stigma to Style: Designing Personalized and Functional AFOs through 3D Printing

Nain

Table of Contents

1   Introduction

1.1   Background

Ankle-foot orthoses – or short 'AFOs' – are a type of orthosis worn around the lower leg and foot with the aim of supporting the ankle position and movement to compensate for weak control over the ankle mobility. They are one of the most commonly prescribed leg orthoses [1], however, there are limited possibilities to manipulate appearances for customization on top of a long and inefficient production process that has not changed in over 50 years [2, p. 145] [3, p. 162] and wearing one can still be felt like a stigma to users. [4] With advances in Additive Manufacturing (or short: ADM), there are possibilities to improve upon this. Furthermore, there are opportunities to address new aesthetics to better reflect the patient's personal style, implemented in their desired appearance of the device.

The following individual work has been conducted as a book project for graduating a master's programme in Industrial Design at Jönköping University with a Master of Science degree, running for 14 weeks from February to May 2020. For this book project, the student has teamed up together with Jönköping University's School of Health and Welfare who are interested in improving services for patients with disabilities and in the long term to gain patents and attract potential collaboration partners for school. The task was given by the associate dean of research and associate professor in the Department of Rehabilitation Nerrolyn Ramstrand, therefore the product developed is for the life science industry, more specifically for the field of orthotics and rehabilitation healthcare.

1.2   Purpose and research question

The objective is to design an AFO that uses an ADM technique that should bring in new potential which users can classify as more aesthetic compared to the currently available designs. A conceptual design that demonstrates the new looks and its new manufacturing process involving additive manufacturing is to be delivered as part of the results.

Specific research questions to be answered are:

What can ADM as a manufacturing technique offer orthotists and patients in terms of choices and self-expression?

How can an ankle-foot orthosis be made aesthetically pleasing whilst still meeting functional needs?

Is it possible to make them look more aesthetical or personalized so that they reflect each patient's personality?

The target group shall be all patients who regularly use AFOs and will include young children, adults and elderly persons and the perspective to be taken must be global, meaning the approach to solve the questions will be taken from a holistic viewpoint.

1.3   Delimitations

This work will not cover extensive user testing phases with a functional prototype, nor does it include a prototype made from the exact manufacturing method that will be determined to be most efficient for the design. Due to time and costs limitations, this project will be conducted in a more conceptual style. Furthermore, contemporary circumstances that occurred in Spring 2020 worldwide due to a pandemic (COVID-19) put limitations on offline research methods that would have been used otherwise. The prototype delivered will only be an appearance model made from methods available on campus with the goal of communicating aesthetics and style in relation and proportions with a person wearing it. Tests to prove its medical function or to validate the end user's experience with the final concept are not part of this research project.

1.4   Disposition

This report will follow the hereafter presented structure after the current Chapter 1: 'Introduction':

Chapter 2: 'Theoretical Background' presents the theoretical framework topics that are essential for this project, then

Chapter 3: 'Method' lists up which tools, strategies, and research methods will be used in implementation in the next chapters.

Chapter 4: 'Research and Implementation' is the main part of the report, explaining what has been undertaken in order to develop the new AFO design. It is made up in two iteration phases, with the second one starting after a Mid-Presentation and a large scale feedback session and runs until the end of the project.

Chapter 5: 'Result' presents the new AFO concept on an aesthetic and technical level and presents design variations with all its details and exhibits the appearance model and other visual material.

Chapter 6: 'Conclusion and Discussion' reflects on the efficiency of the methods used and in general, evaluates the of the result along with a prognosis on how to progress further with the result.

The report ends after the discussion and a reference list and a figure list is included after all the written chapters. An appendix containing various visual and statistic material related to or used in this project is attached at the end of this file.

2   Theoretical Background

2.1   Industrial Design

Industrial Design is a complex design field that focuses on creating concepts for products that are often intended to be mass-produced and sold to users. It is a profession where the goal is to optimize function, aesthetics or outer appearance, and costs to improve or invent a product that can be manufactured with available methods and be sold and compete on the current or future market, benefiting both the end user and the manufacturer or industry. [5, p. 5]

People whose profession is to be an Industrial Designer have an expertise in implementing perceptions and interpretations, which are stimulated from their sensations mixed with their experiential knowledge, to create ideas and design concepts. That experience-driven knowledge, which takes time to acquire but then cannot be unlearnt, is called 'tacit knowledge' and is the result of living through past sensations. Tacit knowledge is therefore crucial to own as a designer and is gained through training to see good design and to correctly interpret perceived user feedback and feelings [6]. Furthermore, a designer is also an expert in gathering and handling different fields of expertise that were unknown to them prior to the launch of a project. The ability to quickly adapt to different fields and to pinpoint and search for the required information is another skill that has been trained for and makes up a part of a designer's tacit knowledge.

In every design project, an industrial designer will put their tacit knowledge into the result, making it a part of the outcome. That means that the design result is achieved through an accumulation of projected and interpreted sensations and perceptions, resurfaced past experiences that relate to the problem, and practical actions that translate the designer's perception into something tangible. To arrive at that tangible product that can be communicated to others, a very practical approach is needed which is explained in methods in the next chapter. Through doing and practicing design itself, the designer enters a cycle of perceiving, evaluating, thinking, generating and visualizing ideas, and again perceiving, which is why sketching with a pen on paper or prototyping with solid materials for example is a method that embodies all of it.

That same tacit knowledge together with a designer's perception is also used to evaluate and ultimately pick options, which can also be called 'intuition' [7, p. 142]. As every person differs from another, so does every designer differ in style and performance from their colleague depending on what past experiences and knowledges they have acquired. The tacit knowledge that every designer applies in their project is what makes them and their work unique.

2.2 User Centered Design

The fact that all products designed are developed for a user of a certain target group and understanding that a designer has no certain knowledge about that user's thoughts and feelings, creates the basis of this design field. To create a highly usable and applicable product, it is necessary to focus on the end user, his comfort, and life experience. If possible, the user is to be involved with the design process that should consist of gathering first-hand information through conversations or collecting feedback from tests with users. [8, p. 9] Users are judges to validate any concepts and to declare whether a new design concept is good or bad. User-centered design is an iterative process but with a stronger focus on the user's feelings and trusting his judgement as the foundation. [9]

2.3 Design Engineering

As Braun designer Dieter Rams pointed out once, it is necessary for good design to be "innovative" [10], meaning that it shall make use of new technology instead of simply re-inventing the shape and thus, copying an existing product. To fully integrate the idea of innovation through engineering, it is important to apply as much technical and scientific knowledge as possible to pay attention to the interior and mechanism of a product. By researching and analysing current and upcoming technology, manufacturing methods, and new materials, a designer can reflect on how to approach the design process. Breaking these mechanisms or technological components apart, filtering out the useful ones and combining them in a way that it contributes to a concept that enhances the user's comfort and quality of life should be an important aspect of this project. Making use of all resources available with an affinity to integrate recently published ones instead of copying old mechanisms is what makes a design innovative and valuable.

2.4 Design Thinking

To quote Tim Brown, CEO of IDEO, he once said:

> "Design thinking is a human-centered approach to innovation that draws from the designer's toolkit to integrate the needs of people, the possibilities of technology, and the requirements for business success." [11]

It is a methodology used in many design fields, not only product design, for problem solving to innovation by applying human centered methods. When designing new products, it is an experimental practice expressed in several iterative phases that are driven by user needs to improve and innovate. Therefore, the design research conducted for product design is not research about what already exists, but it is research through designing what is ought to be. [7, p. 11]

2.5   Bootcamp Bootleg

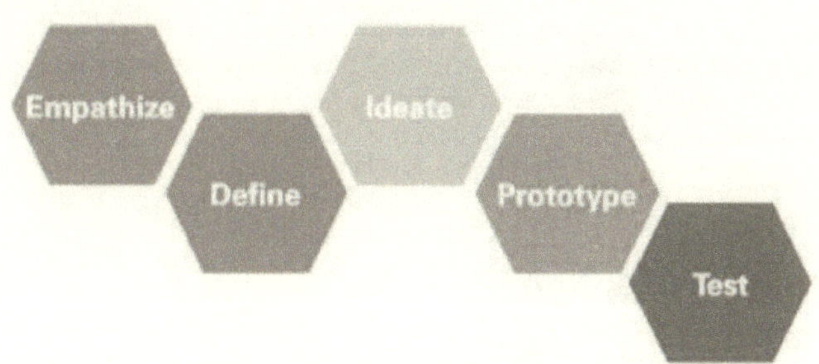

Figure 1, Bootcamp Bootleg Overview [12]

Bootcamp bootleg is a design theory used in design thinking that describes an iterative working process consisting of five phases as seen in Fig.1 that designers will go through from the start to the launch of a new product. [12] Any of these five phases may be iterated again when findings in other phases deem that to be useful or necessary. Bootcamp Bootleg's five phases are as follows:

EMPATHISE – The first phase requires research evolving around the user, the environment, and the product. How are these put into context and what is the viewpoint of the user on the product? What impacts does the product have in both a positive and in a negative way? This human centered design phase will show the real needs and current problems of the user and help in narrowing down the actual target group.
DEFINE – Having understood the user's needs, specific problems and goals must be defined in a way that a result can be judged to be a success or a failure in solving those problems. The problems and questions formulated should represent the user's viewpoint that has been previously researched and must be formulated in a way that it sets an open ground for creative thinking and open solutions as well as inspire brainstorming an ideation.
IDEATE – This phase involves generating a large volume of solution alternatives and part solutions that may evolve, change, be refined or discarded or simply represent a basis for another variety of solutions. Keeping the volume and variety high is key and requires an open and welcoming mindset to all kinds of solutions. All concepts must be narrowed down in accordance with the defined problems and whether they represent the user's viewpoint or not. After an iterative process of creating, assessing, filtering, combining and rejecting concepts, the last ones standing will be considered for prototyping in 2D or 3D in the next phase.

PROTOTYPE – Translating a concept into a tangible and interactive object to communicate purpose and visualize ideas makes a concept accessible for feedback and reflection. Transitioning from a 2-dimensional concept to making a 3-dimensional prototype that reflects the actual measurements in scale may point out new problems to be solved before the prototype is to be completed. Stepping back from the prototype halfway to define and ideate again is a common iterative process and proves to be a useful step back.

TEST – No design will be launched without testing it beforehand, even if it is only a small number of participants. Even simple and short tests might give valuable insight and point out problems and possible dangers that must be addressed and resolved before launching the product to the public. Therefore, it is crucial to involve a testing phase and then iterating all four phases from before if needed, in order to sell a safe and optimized product.

2.6   Design & Emotion

An emotion that a user feels towards a product can greatly influence its success and functionality and is mostly connected to its aesthetic design more than the sole function and purpose of the product. [13] If a product evokes a positive feeling in the user, he will use a product more often and form an emotional attachment to it. Negative feelings or experiences connected to a product may slow down sales or cause a user to neglect using a product, even if it still functions and serves its purpose. An emotional rejection can prevent a user from regularly using an essential product as he can value his feelings more than rational facts, therefore, it is important to research the user's emotions and take them into account when developing a product.

Building a bridge between design and emotions means to understand cognitive ergonomics which is closely related to perception and sensation of both the designer and the user. An industrial designer makes use of their tacit knowledge and skills – like explained in 2.1 but more explicitly – to interpret a design, a mere shape, or any texture to assign an emotion to it, which is then called the design intention. To test if that intention is perceived in the same way by users or the client, it needs to be tested and measured. A measuring tool can be close observation of users, whose actions and reactions must be interpreted by the designer using tacit knowledge. The results are often verbalized with words like adjectives which make it easy to communicate with others that may not be designers. A more specific method to measure is Kansei Engineering, which will be explained in detail in the next chapter. It makes us of words that are ranked by users or the client after how much they match a design intention. That ranking and other methods will then be the criteria to judge the emotional value of a design based on how well a design responds to a design intention.

2.7   Orthotics

In Orthotics, products offered are custom-made or custom-fit wearables to support limbs of patients with neuromuscular and musculoskeletal impairments that make them unable to use their existing limbs like a healthy person. The goal of an orthotist is to rehabilitate people, who suffered from injuries, to their highest level of functional and mobile independence possible to ensure they feel safe and satisfied with their overall quality of life. Orthotists are responsible for evaluating a patient's functional but also cosmetic needs and accompany a patient through the whole process of measuring, selecting, designing, fabricating, and fitting an orthosis. [2, pp. 3, 12, 38]

2.8   Ankle Foot Orthosis

Ankle-foot orthoses – or short 'AFOs' – are a type of orthosis worn around the lower leg and foot with the aim of supporting the ankle position and movement when patients have so called foot drop resulting from weakness in muscles and nerves caused by e.g. paralysis. The AFO supports the foot especially during walking, when the foot is in the air during in the swing phase of gait. [2, p. 225] Without any support from below the foot, it would drop and hinder the patient from landing safely on the ground due to insufficient clearance. It is important to note that AFOs are also the most frequently prescribed lower extremity orthoses [2, p. 225].

A typical AFO is a hard-plastic shell that wraps around the posterior side of the lower leg and embraces the heel and bottom of the foot as shown in Fig. 2 below. It ends not right below the knee but leaves some clearance so the sharp edges do not hit against a main nerve and usually has two straps, one on the top and one around the ankle, for fitting it tightly around the leg and to distribute forces evenly.

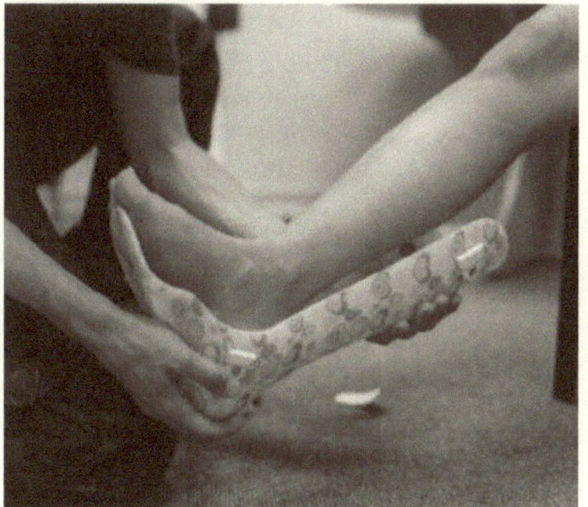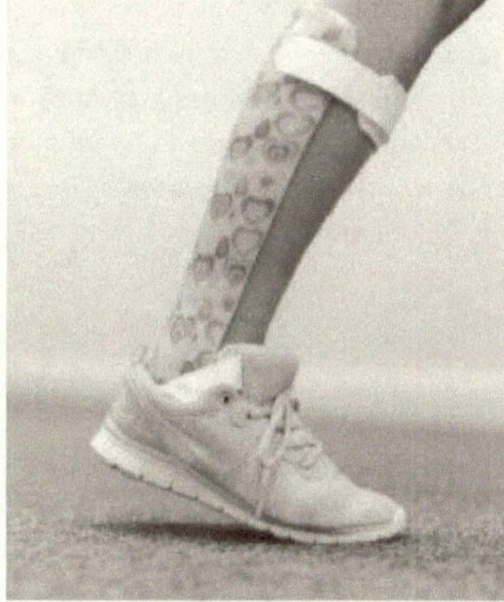

Figure 2, AFO with two straps made from a thermoplastic material with pattern from transfer paper, being donned (left) and worn together with appropriate footwear (right) [14] [15]

There are various types of AFO on the market and variations with only one strap are very common as well as the ones shown below in Fig. 3.

More dynamic and less tight-fitting variations with a spring component exist like shown above in Fig. 3 on the right, however, this project focuses on a static AFO, also called solid AFO or SAFO. [2, p. 225] That is because the client wants to work on an assumption that any results achieved for the shape of a SAFO should be applicable to other types like the dynamic AFOs that are only slightly different in shape but very similar in function and structure.

2.9 Gait

Gait is the manner of walking, while it is considered normal gait when there is enough stability and strength to support the body weight against gravity in order to move forward and walk. A healthy gait also shows mobility of body segments and motor control over them, as well as the stability and control to transfer the body weight from one to another limb while stabilizing the body's centre of mass without much vertical dislocation. Ideally, the centre of mass and the body is moving forward at a person's desired walking speed, stride length and towards their desired direction with minimal vertical and mediolateral displacement.

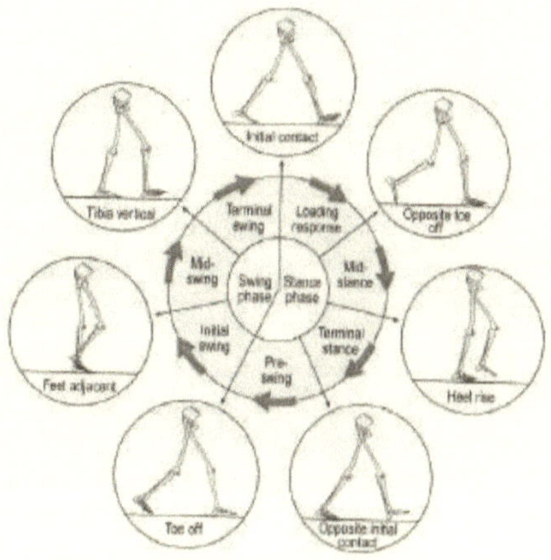

Figure 4, Gait Cycle Overview [3]

However, when someone has an abnormal gait due to impairments, correcting products like an orthosis can be introduced to improve and rehabilitate their gait. [2, pp. 25, 102] A gait cycle is the time interval between two contacts of the same foot with the ground in the repetitive event of walking a certain distance. [2, p. 103] A gait cycle has two phases shown above in Fig. 4, called stance and swing, with the stance one occurring when the foot is in contact with the floor and the swing phase describing the time frame when the foot is in the air. The AFO is there to support a patient especially during swing phase but it also helps in being able to stand firmly during stance phase or when standing and not walking.

2.10 Energy Cost of Walking

For people with impairments, walking without an orthosis means higher energy cost of walking. It is preferred to improve the walking speed and gait kinematics. Walking with an AFO will always reduce energy expenditure, but only when it is the right fit and type, which is the clinician's responsibility to determine upon each patient's deficits. When stroke survivors have their posture and gait symmetry corrected, their walking speed and energy costs are improved, and oxygen consumption drastically reduced. The goal of a treatment plan that includes an ankle-foot orthosis is optimal energy efficiency while walking. [2, p. 28]

2.11 Semantics and Semiotics

Design is also perceived as the aesthetical part of a product where so-called gestalt laws have an influence on the user perception. When designing a product, one of the designer's tasks is to consider how the user will perceive a product. It does only take seconds to make a first impression of an object and decide if it is appealing or not, therefore, it is important to pay attention to visual perception and aesthetics in advance. This arrangement of parts, functions, colours, form and more that make a product a whole is called the Gestalt. [19, p. 296]

Gestalt is a German word and its meaning is related with form, shape, and appearance. The Gestalt psychology is linked to the human brain's nature to find order and structure in everything it sees, searching for patterns and shapes with a view to create an understanding of an object or a product. The most pertinent rules of Gestalt perception shown below in Fig. 5 are:

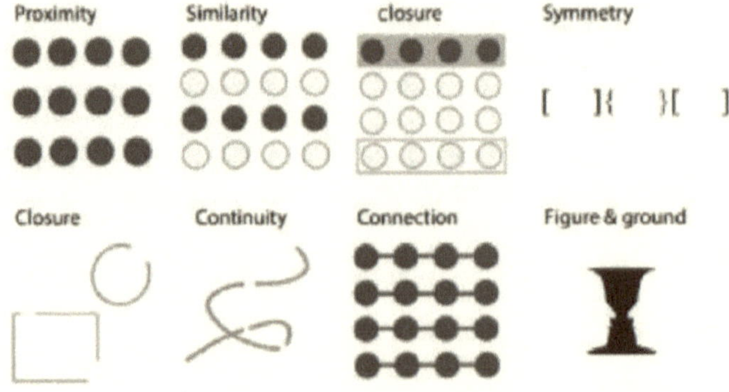

Figure 5, Visuals representing the Gestalt principles

- **Similarity:** Objects that look similar to each other will visually be perceived as a group, or in a structure or a pattern
- **Continuity:** The human brain instinctively looks for relationships between shapes and lines beyond their ending points, trying to extend recognized patterns
- **Closure:** Closure is a phenomenon that happens when the human brain completes open figures on its own accord, filling up the gaps in between
- **Proximity:** When presented with many figures, the shapes that are placed closest to each other are perceived in groups
- **Symmetry:** Two objects that are placed symmetrically to each other will stand out from other shapes and be registered to have a symmetric relationship
- **Connection:** Figures that are visually connected with something as thin as a line are perceived in a group

- **Figure/Ground:** Fore- and background of a high contrast are perceived as figures depending on characteristics such as the colour or size. [20, 19, pp. 297-298]

The gestalt laws are closely related to Semiotics which are the studies of signs in a structured system. Semantics are a subcategory of Semiotics that are the studies of the message and the meaning of the signs. It is the tool for making a product understandable, helping the user to comprehend its features. There are several ways in which the patterns, shapes, or signs of a product's Gestalt can be interpreted. They are categorized as:

- **Describe:** A shape, colour, or texture should describe the product's purpose and its function. It can also suggest the way it should be handled or used.
- **Express:** The form language should express the product properties, which can be robust, lightweight, or fragile for example.
- **Exhort:** The Gestalt of the product can prompt the user to act or react in a certain way before he has gotten a chance to consciously interpret the intended underlying message. An example for a good exhortation is an impulse, which is a reaction that does not require conscious reflection before reaction.
- **Identify:** A product feature can show its origin, affinity, or its belonging in a family that could be for example a brand. Those features are often implemented with icons, logotypes, patterns, or colour combinations. [21, pp. 81-112]

These four functions should be clear to have a good product understanding.

2.12 Ergonomics

Ergonomics – also called Human Factors – is a scientific discipline concerned with how the human being interacts with its environment in any situation working or living. Elements of the environments to consider are the systems, structures, products, organizations, tasks, and interactions with human beings. [22] The goal of any ergonomic studies or project is to optimize human well-being and improve effectiveness and efficiency of work while preventing accidents and minimizing sudden and long-term illnesses caused by work. There are major fields in Ergonomics that will be explained later in Methods along with the appropriate tools to use.

To make a product fit a user, a designer must take physical dimensions and boundaries into consideration. Physical Ergonomics provide or guide how to retrieve data for anthropometric, anatomical, and biomechanical measurements and point out postures and sizes that are neutral, good, and bad for working with certain environments, tools, duration, and repetitive movements. [22] Such data can be retrieved from tables and software must be used with caution due to credibility and changes over time. Applying anthropometric data can help in making the use of a product comfortable or suggesting a working position that is putting the least stress on the body.

2.13 Additive Manufacturing

Additive Manufacturing, also abbreviated as 'ADM', is a more recent category of manufacturing methods where the machine produces the product from a computer aided design (CAD) file by building layers upon layers, hence the name 'additive', without any supportive tools needed. [23] Fig. 6 below shows an example of a product made with the so-called FDM (Fusion Deposition Modelling) method that produces thin layers, which are visible on the image, of molten plastic material, extruded from a nozzle that draws the desired shape in its movement.

Figure 6, 3D printed bolt, built up in layers with FDM method [24]

ADM has an ever growing and developing market and is also commonly known as '3D-printing' or 'rapid prototyping', even though the latter is a subset of additive manufacturing only. There are many different methods in ADM to connect the layers together and it is possible to use almost any material to make the product with, including plastics, metals, composite materials even human tissue. [25] The raw material used for printing comes in various forms like filament or powder and can be fused together physically, thermally, or chemically. Advantages of using ADM as the manufacturing method is that it comes with new possibilities to produce geometries or products in only one step that would not be possible otherwise with traditional manufacturing methods.

3 Method

3.1 GANTT schedule

In order to successfully conduct a project, time and task management has to be carried out and making a GANTT schedule helps distributing tasks and planning ahead to be ready for deadlines and presentations. This project management tool consists of a horizontal bar chart that visualizes all design phases among the available project time to distribute the predicted workload in a realistic scatter. The actual execution and predetermined durations are subject to change as the GANTT chart is mainly used as a planning and tracking tool. [26, p. 131] For this project the planning and mapping tool on the website Team Gannt on www.teamgantt.net is used.

3.2 Literature Review

One of the simplest ways to obtain information and research topics is through reading published literature in the form of books, textbooks, or peer-reviewed articles that are obtainable through libraries online and offline. Learning how to search for the right literature by using correct keywords and using filters can ease the process of finding the right material to read. [7, p. 25]

3.3 Internet Searches

The internet has become a vast place where information is available at any time in almost endless quantities and while acquired information is often not from reliable sources, it is a research method that is convenient and more accessible than books in libraries, even if most of them are available online as well nowadays. Whenever a literature review is not good enough of a method anymore or information that is not printed in books is needed, internet searches can give quick answers to questions and even deliver visual material. [7, p. 51] Internet searches are conducted using tools called search engines. The search engine of choice for this project is Google.

3.4 Function Analysis

To break down all functions of a product, they will be sorted into main, sub and optional support functions, described by a verb and a noun as the object. The description must use clear verbs and nouns and the result must be somehow measurable with the analysis, so it can be validated if the functions have been met or not. [26, pp. 177-178] The function analysis that lists functions can also be used as a checklist for product design specification evaluation at the end of the design process. [7, pp. 139-140]

3.5 Perceptual Mapping

Maps are visually appealing and mostly self-explanatory to the spectator and therefore a good tool to summarize findings or comparisons. By sorting items of the same category in a grid made up of a vertical and a horizontal axis that each indicate a high and low value of a certain product quality, one can see where similarities and overlapping exist but also openings and opportunities for new combinations. A perceptual map is especially useful when doing market research with all competitors and brands that offer the same product as the subject of a project. [7, pp. 80-81]

3.6 Market Research

To get even closer to a product and empathize with the user, it is advisable to go on a trip or go online to see the actual product being sold, purchasing and using them, pretending to be the end user. By doing this, the designer's perception of the products will change, and he will get a deeper understanding of it. Using and testing the product himself once they are purchased is another way to get to know the product rapidly. [7, pp. 76-77] If that is not possible for whatever reason or the product is hard to obtain, a review of products online to spot competitors is enough and will serve well for a competitor analysis.

3.7 Trend Spotting

Milton defines a trend as 'something that has already begun, and trends are therefore spotted rather than created'. [7, p. 31] Once a trend is spotted, it can be analysed in detail and its characteristics and qualities processed later for ideating concepts. Trends can be of commercial or visual nature or there can be also trends in design in general. Identifying trends helps to react accordingly to market opportunities and the economy.

3.8 Persona

Personas are profiles of fictional people who represent the target group, relevant to the project and product. By creating personas, a designer can have a clearer image of the end user by personifying him, deciding on his needs, frustrations, and background. The product can then be designed around a persona and if it fits them, it will also fit the rest of the target group. It is best to create two or three personas per project, representing different genders, age groups and backgrounds or profession that are in the target group. [7, pp. 80-82]

3.9   Scenarios

Having written some personas, it is advisable to put them into possible situations where they interact with the product that is subject of the project, to find out what questions or issues could come up. From those issues, a designer can come up with solutions and a scenario is a great story telling tool to communicate why the solution is needed. [7, p. 32]

3.10   Role playing

For a designer to empathize even more with an end user, especially if it is a very specific kind of target group, difficult to empathize with, role playing or living in their situation for a while is an effective method to gain the same first-hand experience as the target audience. [7, p. 33]

3.11   User Observation

As a part of doing User Centered Design, it is important to get to know the user and how he interacts and behaves with a task or product at hand being in his natural environment. The easiest and most efficient way to star is to simply observe the user without interrupting or disturbing him by staying in the background and keeping a certain distance. Taking photos or recording video material is recommended for further analysis and proof. By letting the user work and act as they would usually work, the observer can see the most obvious struggles the user has and all the errors he makes that may be due to unfitting design. Observing also helps with empathising with the user and gaining better understanding of the work environment and sometimes reveals details that the user would not talk about in interviews because it happens subconsciously. [7, p. 85]

3.12   User Interview & Expert Interview

User interviews are an essential method in User Centered Design; they are conducted to determine the target group's current problems, needs, and wishes by talking directly to its users. It is important to prepare the right questions before an interview so the right information is extracted, and it is recommended to ask open questions so the user can talk freely. [7, p. 70] Similar to interviewing the consumer, it is also advisable to interview experts of field of study. The advantage is the possibility to gain a large of amount of knowledge in a short time, cutting down the amount of research done on one's own through multiple sources, but an interview does not substitute further primary research. [27] Expert interviews offer deeper insight into the manufacturing and engineering environment of the product, how the current market situation is, recent innovations but sometimes also user behaviour patterns that is hard to analyse without a big number of single users to interview or observe. [8, p. 43]

3.13 Questionnaires

A questionnaire is used to collect data from users that are presented with a list of
questions that are sometimes paired with images of things they are asked about. [7, p. 69]
It is a simple yet effective way to obtain information and can be distributed in various
media formats like written and printed or published online. The only disadvantage is that
the handed in results are the only forms of communication between the testing person
and the designer, when 90 percent of the communication is visual in forms of gestures for
example, but those aspects are lost. By publishing a questionnaire online, it is easy to
reach out to a bigger number of people and it can save time by potentially being able to
ask different people simultaneously.

3.14 Mood Boards

Image boards or collages made from 3-5 or more images to convey a certain feeling, style
or brand are a great visual help to communicate the expectations or desired aesthetics for
the upcoming concepts. [7, p. 78] A higher number of pictures does not necessarily lead
to a better result, however, the message within one mood board should be clear and not
misleading. They are usually created between the Research and Concepting phase or as
part of DEFINE in Bootcamp Bootleg. Mood boards can be used in an iteration after the
IDEATION phase to check whether the concepts still match with a set mood and style
or not. Combining pictures with a few keywords on the image board is recommended so
that an outside spectator/reader can combine both to understand the intended mood.

3.15 Brainstorming

Brainstorming is a method to generate a very large volume of ideas in a short amount of
time. All ideas shall be recorded or put down either in drawings or words and the aim is
to think as widely as possible while disregarding all limitations for now. Therefore, no
criticism and filtering should be presented to enable the flow of creativity and end up with
as many concepts and ideas as possible. [26, p. 190]

3.16 Mind Mapping & Word Cloud

To visually present valuable ideas from brainstorming or just to connect ideas, mind maps
come in handy since they use words and arrange them in a hierarchy by emphasizing head
themes or important ides by highlighting, encircling, or simply writing them in bigger
sizes. Different ideas can be connected with lines; therefore, mind maps encourage
thinking out of the box instead of in a linear way and also make good visual presentation
material that communicates in a simple and straightforward way. [7, pp. 56,66]

3.17 Sketch Ideation

Like in brainstorming, it is important to document the ideation process and sketches are a powerful tool to communicate quicker than by using words that has been practiced by designers and architects for decades. [7, pp. 34-35] In this way, many alternatives are created and documented and in case the concept needs to change again, the designer has resources available and can get back at sketches that were documented. In sketch ideation, the drawings can be made quickly and rough but if needed, sketches can also be cleaned and coloured, all depending on what the designer wants to communicate and memorize with the sketches.

3.18 Silhouette Thumbnails

In case that Sketch Ideation exhausts the currently available creative output, drawing silhouettes for ideation will be an alternative. It is another method used by character designers and finds application in other design fields such as Industrial Design as well. It is another method for creating a huge quantity of ideas in a short period. [28] Instead of line drawings, the complete surface area of a product in a side-view will be painted, not sketched. Because only the silhouette will be drawn, the different concepts are bound to be distinctive from each other only from their shadows, allowing the designer to come up with a variety of unique ideas. Once the silhouette has been evaluated as having potential, the designer will start sketching on top of it to define further shapes and split lines going from the outsides of the sketch to the inside. [29]

3.19 Digital Graphics & Vector Graphics

For some concepts it is necessary to create scalable graphics or curves, also called vector graphics. Those can be used for further ideation in 3D modelling software or for printing purposes and are created in either Adobe Illustrator or Inkscape. Inkscape is mainly used to convert a non-scalable pixel image to pixel graphics which saves time for complex drawings or patterns. The process of making digital graphics is the same as in 3.15 'Sketch Ideation' [7, pp. 34-35] and can use silhouette thumbnails as well [28] [29] but it is using only digital tools and it is possible to spend more time on selected designs to prepare them for the prototyping phase. The principles from semantics and semiotics will be applied when creating graphics that only communicate with two shades like black and white.

3.20 Kansei Engineering & Focus Group

Kansei is a Japanese word that has no direct translation to English, but it is describing feeling, emotion, and impression, and describes the emotion that a user has as a response upon perceiving a product [30]. After making a first impression and a judgement, that product will continue evoking the same emotion – negative or positive – in the user for a lifetime.

A method to connect design with emotions and to test whether they match up with the design intention is to use describing words – called Kansei words – as suggested by Kansei Engineering methodology. Those words can be asked from a (potential) target group users in a focus group session where they are presented with the designs. The Kansei Engineering method is used to test various designs in a conceptual state to decide on the final concept [30].

A focus group is usually made up of at least 3-10 people and is a smaller size group for holding sessions involving user centered methods to observe, interview, and test [7, pp. 70-71]. The focus group can also be presented with the designs and the words in combination and users are asked to rate whether they fit or not, or to rank the designs according to their personal preferences, optimally, discussions will reveal their own Kansei words to describe why they like or dislike. Using words creates something tangible to document, so there is a tool to map and rank concepts, which will help sorting the designs by the positive or negative feelings they evoked in a user [30].

3.21 CAD – Computer Aided Design

Computer Aided Design – short 'CAD' – also called 3D-Modelling, is a tool necessary for making prototypes but it is also a playground for digital ideation. CAD software allows the designer to experiment with measurements and shapes as well as colours and proportions when considering digital renderings before the design is translated from 2D into 3D. [26, pp. 118-119] Two programs are used for solid and surface modelling. Rhinoceros 3D will be mainly used because of its ability to manipulate more organic surfaces like the shapes of the human body. It works in an intuitive way that is difficult to achieve in solid modelling. Any closed-up surfaces in Rhinoceros can be exported as solid bodies for finishing treatment like filleting in SolidWorks 2019 if necessary.

3.22 Digital Rendering

Digital Rendering is a form of drawing similar to Sketch Ideation but on a more detailed and elaborate level. A chosen design concept is drawn with digital tools to a degree of detail that it looks photorealistic and creates an illusion of having a finished and real product. Because the images look so realistic, so-called renderings can also be a kind of digital prototype or mock-up but quicker and cost-efficient in a stage where it is not necessary to test with a physical model yet [7, pp. 96-98]. To create such photorealistic presentation images, they can either be drawn by hand or be rendered with a software if there is a 3D model (see 3.21 Digital Simulation). For drawn renderings, digital software is used like Adobe Photoshop, Autodesk Sketchbook, and Clip Studio Paint, that enable the designer to create a photo montage or to use material textures as well as other functions that would not be possible with traditional tools.

3.23 Digital Simulation

Designing a product for a certain target group means it will be used in a specific environment or setting of the user as well. When testing with a working prototype is not possible, computer aided simulations in the CAD programs like SolidWorks and Keyshot can offer help. [7, p. 126] Simulation is a form of prototyping and offers testing for aesthetics and the user's acceptance for it before deciding on the final design. Renderings in Keyshot can help in deciding on materials and colours for aesthetic purposes when the 3D model is already made with CAD. For this project, digital simulation will be used to generate rendering images for presentation purposes since the project's goal is to create a conceptual design.

3.24 Rapid Prototyping & Mock-Ups

When still in the ideation phase, rapid prototyping can be used to confirm or reject digital 3D or 2D concepts. Common methods are additive manufacturing or quick foam models. It is not necessary to build the whole product, but it is often enough to test critical parts like handles and connection mechanism to test functionality and size. Rapid prototyping prevents major errors such as completely wrong sizes or not working constructions by encouraging adjustments to the design before the deadline [7, pp. 107-109]. Simulations in Keyshot can count as mock-ups for colours, texture, and forms as well.

3.25 Appearance Model

When working prototypes are not needed, appearance models can be enough. They communicate size, volume, colour, surface quality, material, texture, weight, and haptic, and offer the client and designer to see the concept in context and move it around to inspect it from every angle. For appearance models, it is important to make it look realistic on the surface, while the appearance of the insides is of no importance and functions such as simple (re-)movable parts are desired but not mandatory. Materials used can be any that the designer think is useful and handy for the product. It is not necessary to use the same material as it would be used in the final retail product but if possible, it is desired to use them [7, pp. 102-104].

3.26 Feedback Session

To narrow all concepts down and decide on the final details, feedback is required. Asking the project supervisor and the client, as well as other designers for their opinions will broaden the designer's horizon with understandings from other point of views [7, pp. 141-142]. After the planned Mid-Presentation for this project, three final concepts will be evaluated, and the teachers and the client will give advice on which concept to pick for the final concept and more feedback is provided weekly with the project supervisor.

4 Approach and Implementation

4.1 EMPATHIZE

The goal of empathizing is to gain knowledge about the product, the user, the industry, technical aspects, requirements to stick to, and through that it is possible to crystalize problem areas to solve which will be presented in the next chapter 4.2 DEFINE.

4.1.1 GANTT schedule

A first step to start with a project is to summarize all tasks that need to be accomplished to reach results in each given time frame. By creating a GANTT schedule, one can plan and gains a visual overview over all tasks. The first draft of the GANTT schedule before the project has started is shown below in Fig.7. A bigger size of a more detailed GANTT schedule is attached in the appendix.

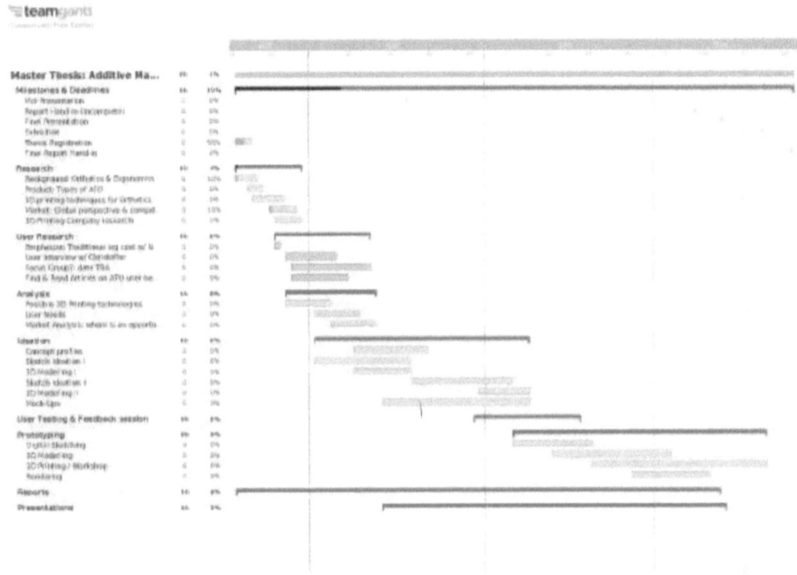

Figure 7, GANTT schedule first draft (minimized view)

First, deadlines and presentation dates are put it as milestones; they set the limit and split the project into timeframes. The tasks to fulfil the project goals are split up using the Bootcamp Bootleg model, separating the phases into empathise (different research phases), define (analysis), ideate, prototype, and test with the addition of separate categories for report and presentation deadlines.

Important dates that were set milestones are:

- the **Mid-Presentation** on 11th March 2020
- the **Final Presentation** of this project on 13th May 2020
- the **report draft hand-in** for the opponent reader on 8th May 2020
- the **final report deadline** on 25th May 2020
- the **exhibition day** on 29th May 2020 with prototype and project presentation

The most important dates that split the project into two ideation phases were the Mid-Presentation and the Final Presentation since both required deliverables that represent concept ideas or the material for the final concept, while examiners and clients would be present to evaluate them.

4.1.2   Background Research: Product and Usage

Because this field of orthotics and rehabilitation therapy in general was unknown, it was necessary to learn more about it and more especially about the ankle foot orthosis as a product as it is not something one is confronted with every day. The target group is very specific and experts working as orthotist are specialists. By understanding the working environment and the users better, it will be easier to define the needs of the industry and the user.

The first step to gather information was to read through literature and to skim through the internet when necessary. A recommended book by the client that is also used as a textbook for universities worldwide, gave the best quick overview about the purposes of orthotics and the most necessary functions of an AFO.

The solid type AFO that this project focusses on prohibits movement of the ankle in any plane, thus resulting in a lost in mobility but providing greater stability during stance phase and enough limb clearance during swing phase. However, due to their rigidity, users compromise in smooth transition through rockers of the stance phase and might need appropriate footwear and cushions to compensate and stimulate these transitions. [2, p. 225]

A patient who has difficulties using his ankle muscles and nerves, usually uses an AFO for support in the swing phase of gait when the foot is in the air, being moved from behind his body to the front and whilst kept above ground without touching it. Since there is no support from below the foot, clearance between the foot and the ground is usually guaranteed via ankle movement. Is that not possible, the foot will drop. To ensure there is enough clearance, a patient with foot drop has to lift his leg much higher, resulting in less energy efficient walking and a risk of falling if initial contact with the ground is made in a wrong angle or position. To prevent that, an AFO keeps that foot position lifted and secured. The phases where and AFO user needs support can be seen in the marked areas of Fig. 8 below. [3, pp. 32-35]

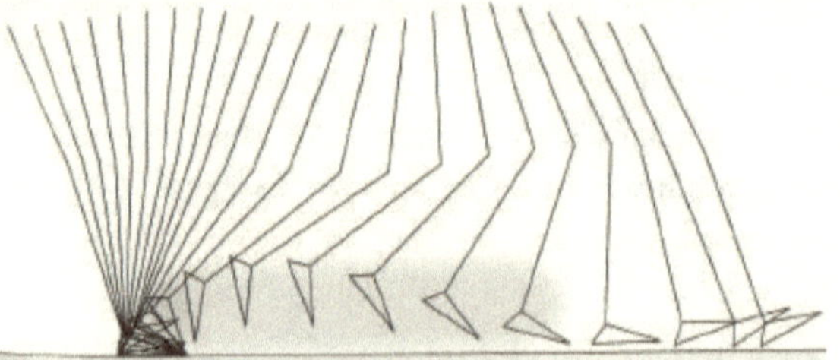

Figure 8, Swing Phase where AFO is needed most (area marked in red)

Further review of many literature resources has shown that there are limited possibilities to manipulate appearances for customization of an AFO, as most images found are showing a similar basic design. It is hinted at that emotional acceptance of the product has an impact on whether the user will use an AFO correctly or often enough for successful rehabilitation and it is strongly connected with cosmesis, the aesthetic aspect in medical healthcare products. [2, p. 354]

4.1.3   Background Research: Current Manufacturing Method

After a better understanding of the product and its use has been established, further research in the field of designing and manufacturing an AFO is needed. From literature review [2, pp. 153-157] and an expert interview, the traditional method to make a custom AFO was a six steps process consisting of the following:

- Taking measurements with measuring tape
- Making a negative cast of the limb
- Making a positive 3D cast of the limb
- Modifying the cast to incorporate desired controls and allow clearances for
- Fabricating the orthosis around the positive cast using thermoplastic
- Fitting the orthosis to the patient and making minor adjustments

The average time is 34-36 hours, according to the interviewed expert and includes the time to let the plaster dry. Fig. 9 and Fig. 10 below show the whole process of all the six steps.

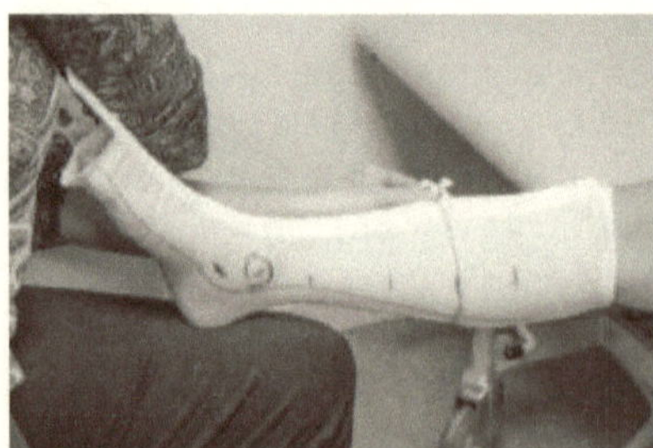
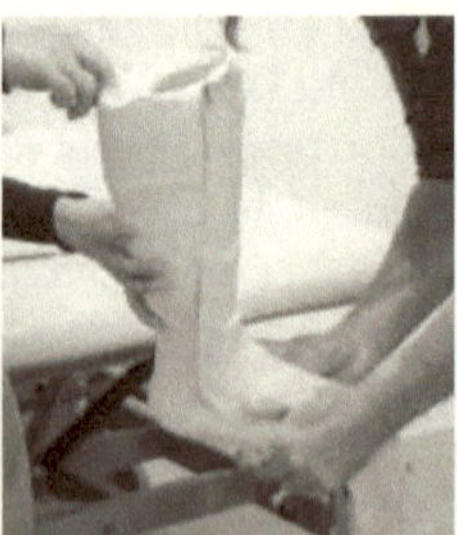

Figure 9, Measuring before casting, fixing negative cast after taking shell off the leg, filling for positive cast

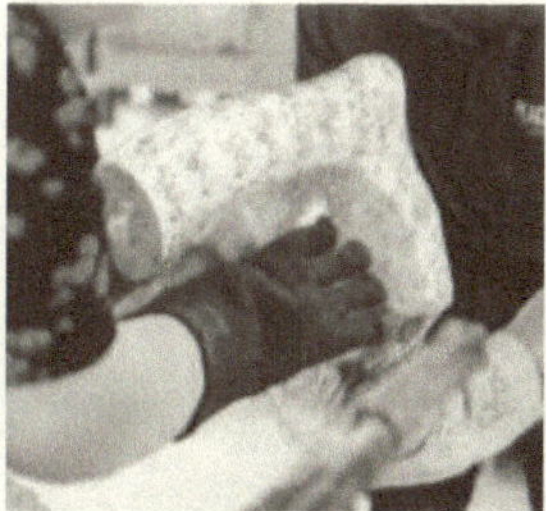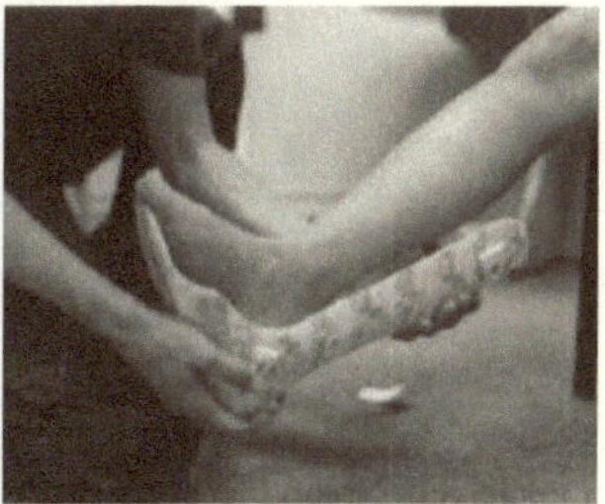

Figure 10, Cast after modification, thermoplastic wrapped over the cast [31], adjustment & fitting to patient [14]

Further literature review showed that first advances in AFO were 50 years ago with the introduction of thermoplastics and 30 years ago with affordable CAD software and computers. [2, p. 145] However, the appearance of AFOs and the basic design has not changed significantly since they were introduced. [3, p. 162]

Apart from the custom method explained before, there are off shelf versions of AFOs that are produced with more efficient mass production methods such as injection moulding, however, they do not offer a fit that is as effective as custom made ones and are only prescribed in less severe cases. Additionally, an attempt to use mass production introducing a modular system is on the market, but experts say that they cannot always replace the handmade custom fitted products because they lack the functionality that some patients with deformities and conditions require. [2, p. 159] Moreover, mass produced AFOs are not subject of this project, the goal is to find a solution for the more common custom-made version.

4.1.4   Background Research: Additive Manufacturing in Orthotics

Although healthcare and insurance companies want to reduce costs when it comes to making orthotics, and there is pressure to be more efficient in the industry, thus competition to find alternative production methods to the labor intensive and expensive traditional method, there has been little advances that involve ADM. [2, p. 159]

One major problem is the use of **CAD software** that seems to be difficult in application because there are many different programs on the market catered to one very specific application in orthotics and prosthetics and owning all these programs and gaining the proficiency costs time, money, and effort that might be hindering practitioners from adapting faster. [Lusardi, p. 161]. All those factors are causing a slow adoption with little results only, making it difficult to receive funding for it, which is why traditional methods are still widely in use.

3D-Scanning however has been used in factories that specialize on orthotic and prosthetic products. They share the same company grounds, saving costs for practitioners and raising the level and consistency of device quality. Such companies have better technicians, resources, and knowledge in manufacturing than a single orthotist can offer. Outsourcing to such companies saves time and effort and the need to have an in-house facility to manufacture orthoses. [2, p. 159] The 3D scan is then with a computer-controlled machine to cut a block of foam into the positive cast shape of the desired leg (CNC milling). The rest of the process is following the traditional method described before in chapter 4.1.3. While 3D scanning saves a significant amount of time and material, those big companies have not successfully implemented 3D Printing yet according to an interview with a digital manufacturing manager [32] and the lack of results when browsing the internet.

To find out more about the use of ADM in Orthotics some **internet research** using keywords like '3d printed' and 'orthosis' has shown results about competitors and private companies and their printing techniques. More about the Market Research can be found in chapter 4.1.7. In combination with reading **peer-reviewed articles**, a summary of the 3D printing techniques used are as follows:

FDM (Fused Deposition Modelling) as the printing technology does not seem to be strong enough to hold all the forces in an AFOs since composed layers have connections that are too brittle and the layers don't hold together very well, but a company called ottobock and Pohlig GmbH in Germany are the only competitors using it so far for orthoses and helmets [33] [34].

Ultimately, the most preferred method is **SLA** (Stereolithography) **or SLS** (Selective Laser Sintering), where the material is melted or hardened to form one solid body [35]. A German company called OT4 uses Jet Fusion by HP [36] which is a SLS method that also creates technical opportunities to apply small hole patterns to make their orthoses breathable. The same printer is also used by competitors like Pohlig GmbH [37], UNYQ [38] and Crispin Orthotics [39].

HP promises that **Jet Fusion** is much more time efficient and achieves a higher quality in finish like smoothness on the surface. By using different agents HP offers, it is possible to control the colour and translucence for aesthetics, stiffness or elasticity for functionality, and surface texture to manipulate wear or friction [40]. It is even possible to control the material printed in a certain spot. All of that can be influenced in steps of the smallest printable unit called voxel.

In general, the few results produced so far are not as strong as traditionally manufactured ones, however, variations in thickness during printing helped solving stability issues [35] but finding the right method would require more testing, which however is in need of funding.

4.1.5 Background Research: Material

Right now, **plastics** are very popular in Orthotics when they are used traditionally, being moulded over a plaster cast, therefore thermoplastics are a great choice, also because minor adjustments can always be done through reheating the material. Popular materials are acrylic, copolymer, PE, PP, PS, and vinyl. Lower extremity orthoses need higher density PE instead of PP. PE has a long fatigue life and is therefore often used for joints, hinges, and shells [2, p. 148].

Carbon fibre composites are often used for their higher strength energy return characteristics [2, p. 28] than steel while being much lighter and also stiffer, however, its impact strength is low, which is why it needs to be reinforced with fibres like Kevlar or fiberglass [2, p. 145].

Foams are used as a protective layer in the shape of paddings in orthoses, protecting vulnerable parts of the skin to pressure or parts that have much friction with the device like bony prominences, however, but foams can also act as an insulator, allowing the product to become very hot when worn for a long time or in warm environments.

Leather in straps or soles has advantages of protecting skin from irritation and it is water permeable, thus, permitting perspiration. Due to its breathability it is preferred over synthetic alternatives to make belts, straps, and cuffs, and can easily be moulded over plastics or metals to create a combination sandwich material. However, for AFOs, **Velcro** straps are more common than leather because they allow a secure fit that is seamless but flexible adjustment in sizes.

The strength of a material is especially important for lower limb devices and density plays a big role because the goal is to make a product to be as lightweight as possible, however that can be at the cost of strength so a compromise has to be found and the perfect balance to be determined. Materials used for orthoses often retain heat and make perspiration difficult. Lower extremity devices might interact with urine due to incontinence, therefore, generally easier to clean materials are preferred [2, p. 146].

Early attempts in 3D printing were lacking in proper strength and reliability compared to traditional orthoses and therefore advances in applying ADM have been slowed down. Those early prototypes used materials such as carbon fibre infused plastics, and filaments out of polycarbonate (PC), nylon (PA), ABS, PLA, PP, TPEs, TPU, PETG, PVA, and ASA [2, p. 161].

Popular materials for printing found out via competitors using internet searches are **polyamide materials/nylons** such as PA11 or PA12 [37] [41] [42] [43]. Another competitor, UNYQ, is also using a polyamide but did not disclose with one, but it is very likely to assume that they are using one of the aforementioned ones as well [38]. A middleman for 3D prints called Shapeways offers PA11 as a printable material that is also bio-degradable [44].

4.1.6   User Research

An attempt at empathizing with the end user was to interview them and experts in the field of Orthotics as well as to observe them while conducting the interview. Finding AFO users specifically for an online survey would have taken too much time or the amount of responses might have not given enough specific data and a lot of visual communication would have been lost.

Fortunately, it was possible to interview two users, one who is an end user who has been wearing AFOs for many years and another expert who is a former orthotist who now works in product development. All interviews have been conducted with a set of prepared questions, and answers were noted down as the interview progressed using a laptop. The specific problems to solve or ideas to implement for the new design are presented in the chapter 4.2 DEFINE.

The first interviewee was **the end user** who was sharing a lot of personal and emotional information and expressed his frustrations and wishes well while telling a life story. Talking one to one like that made it easy to sympathise with the interviewee and the shared experiences are highly valuable information.

Interesting things that were pointed out are:

- The effect of seasons, weather, and heat when wearing an AFO, especially in Summer season, when the padding adds extra layers and non-breathable material makes wearing an AFO an uncomfortable experience
- Differentiating between patients who are born with an impairment and patients who wear an AFO because of an accident or a disease that happened later in life
- Replacing a broken AFO is a common act that is rarely talked about, all AFOs will break at some point but how often depends on the user, his lifestyle, and the forces put on them
- It can take years to find the right type of AFO for a patient, it is a trial and error process

The second interview is with an expert who is a **former orthotist** with 9 years of experience who is now working in product development. This interview was also recorded and transcribed with the interviewee's consent, using Google recorder. Having worked with many patients of all age groups for years, an orthotist is an expert that can relay information about the end users but from a different viewpoint. The orthotist who is fitting the AFO to a patient is often disregarded, even though they are also an indirect user of the product.

Important things to note are:

- an orthotist also carries an emotional load when prescribing an AFO and desires to offer a patient the best solution, having the long-term goal of rehabilitation in mind and wanting the patient to like the product they offer

- elderly people tend to have a focus on functionality and will use the AFO if it shows improvement
- teenagers are the age group with the highest difficulty in accepting and wearing an AFO due to emotional rejection
- the ability to preview the looks of an AFO to the patient is a desirable feature for the future, so disappointments can be avoided through proper communication

4.1.7   Market Research

To get to know the direct competitors and the current style of AFO products offered on the market, an online searches were conducted, looking up brands and companies that were recommended by the interviewed expert but also looking for other competitors using keywords like 'Ankle Foot Orthosis' and '3D printed' in a search engine. Some competitors found were central fabrication companies like Team Olmed [45] that specialize in orthoses and prosthetics and manufacture many different ones in the same company grounds, which saves costs for practitioners and raises the level and consistency of device quality. [2, p. 159] Such companies have better technicians, resources, and knowledge in manufacturing. A selection of **traditionally produced AFOs** including ones made in local clinics, AFOs made in central fabrication companies, and off the shelf products, than can be purchased online without a custom fit, are shown in an image board below in Fig. 11.

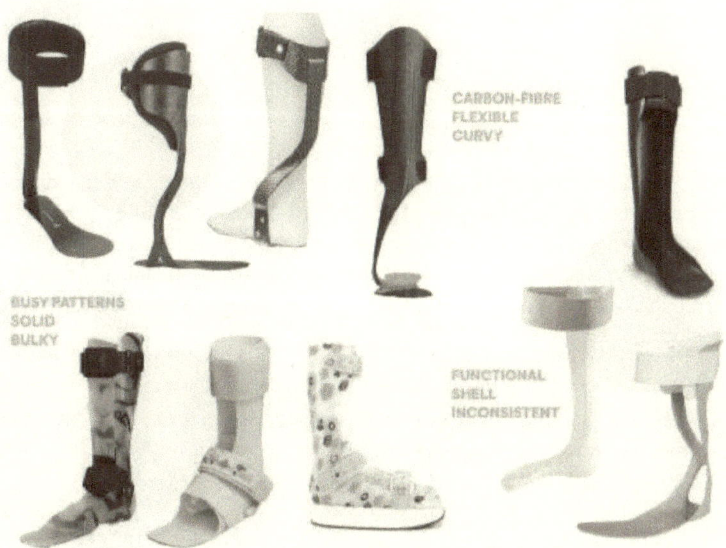

Figure 11, Traditional AFO designs

Further research has led to designs that were produced with ADM technology, however not all competitors' products were AFOs, but also include other kinds of orthoses, mostly for the wrist. It was striking that the few companies that are competing aesthetically and with a salespoint of personalization are all working for prosthetics and not orthotics [38] [46]. Some examples of **3D printed orthotic and prosthetic products** are compiled and shown below in Fig. 12 where the specific AFO designs mostly show simple perforations and hole patterns while the prosthetic product examples demonstrate graphic patterns made from negative spaces on the prothesis' surface.

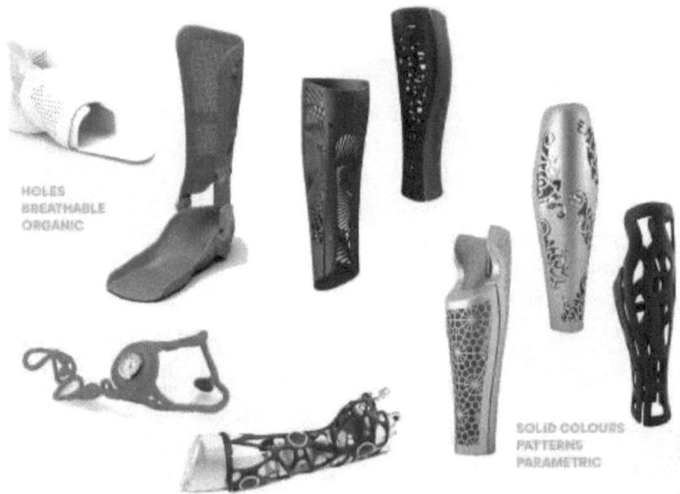

Figure 12, Orthotic and prosthetic products made with ADM technology

To gain more information about the industry, another expert interview with a **digital manufacturing manager**, organising the production of AFOs in one of Scandinavia's biggest suppliers for orthotics, Team Olmed, has been conducted via a phone call. Not many numbers or specific information were disclosed due to confidentiality barriers, however, the interview gave a good first-hand overview over what is relevant for the industry, which are as follows:

- AFOs make up an important amount of sales among all orthoses
- 3D scanning technology is already implemented; scans are used to CNC mill out a positive cast, skipping the plaster cast process
- No 3D printing is implemented in the production process yet

The result of this interview is that AFOs are a significant product for the industry and implementing ADM is a desired feature but also a difficult one that is not well explored yet that they are safe to use. The interviewee showed significant interest in implementing ADM techniques, but it seems that there is a lack of an industrial design department in companies that make orthoses. The only ADM related technology implemented is 3D scanning of the leg, but it is efficient enough to be a regular part of the process already and saves a significant amount of time, material, and errors.

4.1.8   Role Playing

To obtain a better understanding of the traditional process as well as to empathize with the patients that receive an AFO prescription, the designer tried to become the patient and live through their routine when they have a custom AFO fitted to them. The role-playing did not last for a whole day and a scenario where an AFO was worn was impossible because a fitting product was not available and making one would take too long and would require too much assistance. However, they underwent the measuring and casting process that is necessary for ordering a custom AFO. The casting process consists of making the negative cast and then the positive cast. The process is shown below in the images of Fig. 13 and Fig. 14. The casting process was done by an orthotist who is also the client, Nerrolyn Ramstrand, who was willing to help.

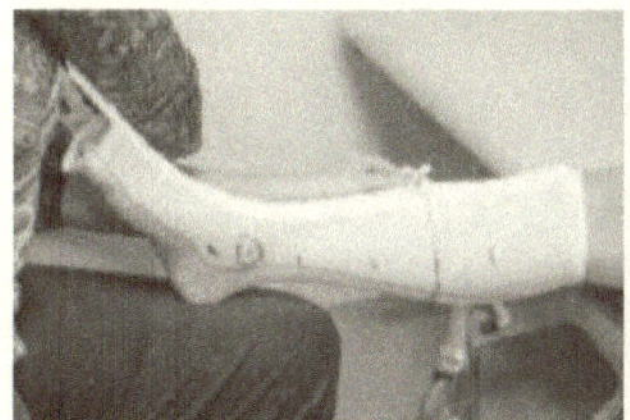
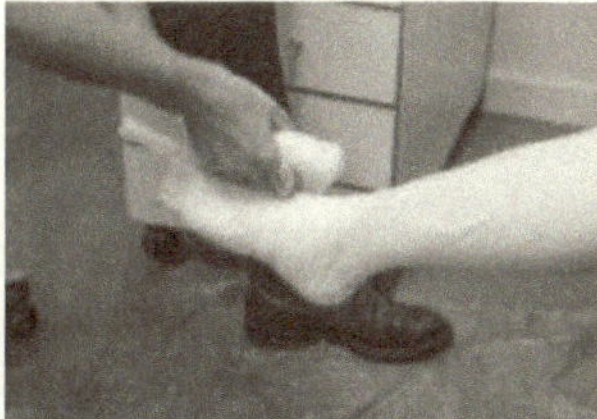

Figure 13, Measuring circumferences and marking areas to be modified (left) and wrapping plaster around the leg (right)

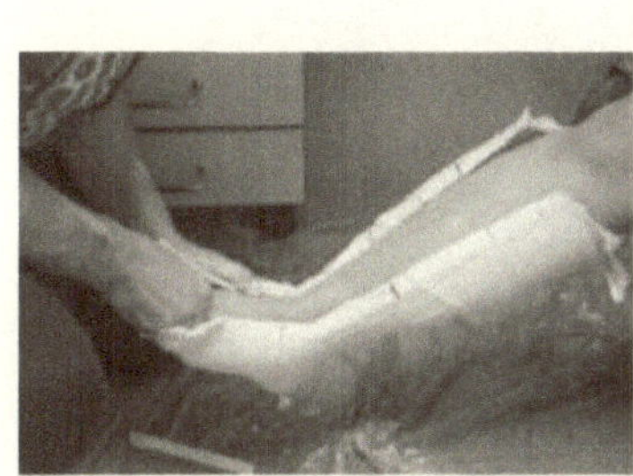
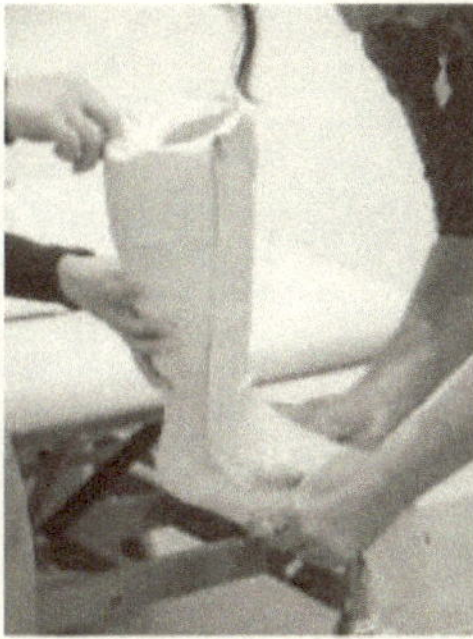

Figure 14, Taking the negative cast off (left), closing up negative cast (middle) and filling with plaster to make positive cast (right)

The negative cast was solely used to produce the positive cast, which will not be used to mould a sheet of plastics over it to make an AFO but for another purpose later in ideation phase, when 3D scanning and 3D modelling is relevant. Because the plaster needed time to dry and its modification needed to be finished by the client, the process was split into several days.

The role-playing has shown that including the drying time for the plaster, the process of making a cast is time consuming and uses up a large amount of material. The cast itself would serve a single time use for wrapping the plastic material around it only, unless another AFO as a replacement with the same dimensions is wanted. Furthermore, a patient who is undergoing this procedure for the first time has no other choice but to trust the orthotist as the cast making process is very specialized. The patient's role is quite passive and there is little room for them to involve themselves in the design process and express his wishes in the cast making phase.

The obtained cast with all its modifications is shown below in Fig. 15 and serves another purpose, which is having a model to scan, so the 3D scan can act as a frame for further CAD modelling. Advantages of taking the designer's leg measurement for the cast instead of those of another test person are the gain of their own experiences as a gateway for empathizing and the future ability to test prototypes on themselves since the measurements should match up.

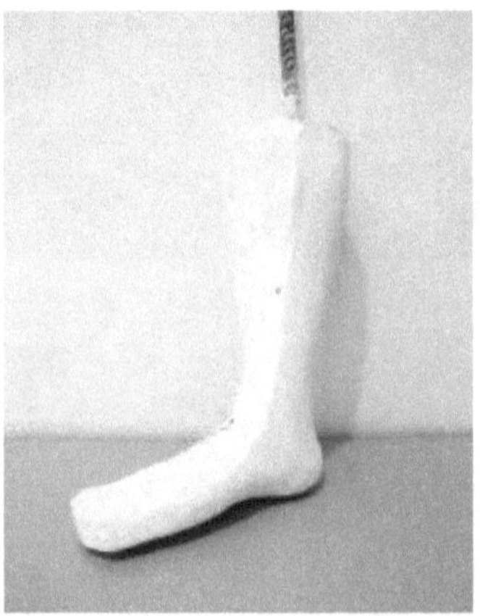

Figure 15, Finished positive leg cast with all modifications (expanded spot for ankle area and elevation for a 1cm high heel)

4.1.9   Personas

Personas were written with the intention to create fictional characters to empathize with that cover all different age groups that were to be considered, with a variation on genders as well. Four personas were created and represent different ages, genders, and backgrounds that made them wear an AFO, living in different countries all over the world. The contents of the personas were influenced and inspired by results from previous user research and the initial problem description. All personas were kept simple and compact on one page only for making them easier to memorize, crystalizing specific needs and frustrations depending on the backstory. An example of a persona is shown below in Fig. 16. Larger versions of all four personas are attached in the appendix.

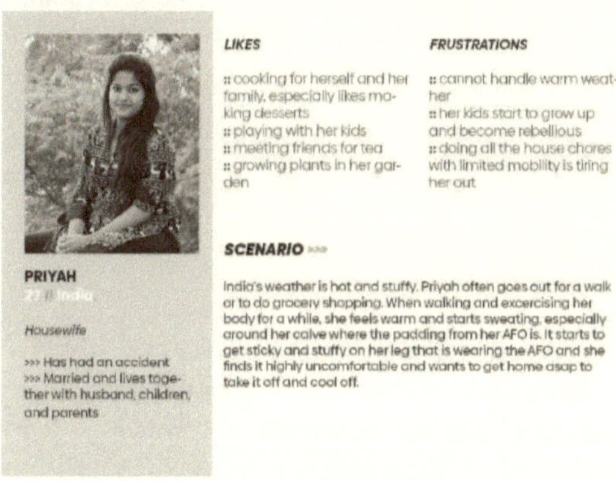

Figure 16, Persona example

The information included on each persona sheet include a profile with their picture, name, age, country of residence, and family situation, followed by insights into their frustrations with the AFO as a product and personal preferences or favourite things they do in their free time. Those insights give them a deeper personality and inspire to think about situations users could be in while wearing the product.

4.1.10 Scenarios

To make better use of the personas, short scenarios were written for each of them on the same page as the persona itself as shown in Fig. 14 in the last paragraph. Each scenario describes a different setting where a certain problem from the user environment and his frustrations that is thought to be typical for their age group.

The settings chosen for the scenarios were based on the following themes, inspired from the user interviews and internet searches:

- Hot weather in warm countries or summer season
- Troubles of finding clothes and footwear to fit the AFO
- Requiring frequent replacements due to growing up & constant wear
- Bullying in groups as a result if stigmatization, especially among teenagers

4.2   DEFINE

This chapter defines which specific problems and needs exist and points out all requirements to be taken into consideration when entering the ideation phase afterwards.

4.2.1   User needs

Interviews with end users and experts in the medical field as well as internet searches have shown, what a current AFO design lacks, which are:

- durability
- air permeability
- self-expression options or having a choice in general
- convenience to reproduce or adjust measurements without a long waiting time and maybe without consulting a doctor even

Different age groups have different understandings about the AFO as a product and therefore different difficulties in accepting the product as an aid that they benefit from rather than a socially negative stigma. An expert interview has grouped users into infants and children, teenagers, young adults and above, and elderly people. Teenagers have the biggest issue with accepting and integrating an AFO into their daily life, but elderly people can also be very stubborn about their habit and tastes, ending up not using the AFO often enough.

A clinical study conducted with 10 patients aged four to ten has shown that all patients are using their AFO less than the prescribed and recommended time [47] which leads to less effect in rehabilitation. Therefore, creating better acceptance is an urgent user need to work on. An AFO being neglected and segregated from a patient's daily life due to its difficulty to wear – physically or emotionally – happens often and defeats the purpose of the product at its very root that is rehabilitation and returning the freedom of mobility to the patient as much as possible. [2, p. 162] To make the patient's process of accepting their device better, the new concept needs to include the following factors as well:

- Better cosmesis factor: aesthetic look that is emotionally easier to wear with fashion items (aesthetics component)
- Easy donning/doffing of the AFO (usability component)
- Easy to clean (maintenance component)

An interviewed user has confirmed that the process of putting an AFO on is quite simple, intuitive, and quick already so that should in no way be more complicated. If possible, further simplification is desired but not required.

Furthermore, patients should have the **choice** if they want to stick out with their AFO or not. If they would rather want to hide it, they should have the choice and means to hide them under clothing as much as possible. The general aesthetical changes needed are to make them look less bulky and stiff, meaning they should become subtle and elegant so they fit easily under pants without sticking out, is what the interviews and a user article online said. [4]

Another important user need is the desire for **customization** and the current customization options are transfer foils with patterns printed on them, which are formed along with the sheet material of the plastic which causes the pattern to warp and be stretched, decreasing its aesthetic value. Another form of customization performed is the application of sticker labels to the AFO. Such practice may be performed on the new design as well for further personalization, what a user makes of the AFO once it is handed over is left to them. More precise printing of images on a surface within a 3D printing machine is technically possible but it shall not be the focus and only solution of this project.

Experts have expressed in interviews that if patients knew what AFO they received, perhaps communicated through computer simulation, that would ease the process of accepting the new wearable device. Driven by the wish to work closer with their patients, experts like orthotist want to make use of digital tools to prevent disappointment and to ensure **communication** on the same eye level. Leaning more to visual than verbal means can eliminate or minimize the use of medical terms that are difficult to understand for outsiders.

Summarized, further user needs to consider are:

- Freedom of choice how to express their AFO (hiding or showing off)
- Freedom to customize (before and after ordering the AFO)
- A tool or a language for communication, visually and verbally

4.2.2 Market needs

The goal of this project is not to launch a product on the market; however, the market analysis gives proof to why the design proposal is relevant even on an economic level. To spot an opportunity in the market and define what kind of product is needed, it was necessary to analyse the competitors and rank them accordingly. A selection of competitors globally has been sorted into a perceptual map as shown below in Fig. 17.

The map does not include all existing competitors but only presents a selection of them. Since the project is conducted with a global approach, many more companies, and clinics, especially those that produce traditionally or offer off the shelf AFOs, exist. However, all companies that produce orthoses with implementing ADM technologies are shown in the map.

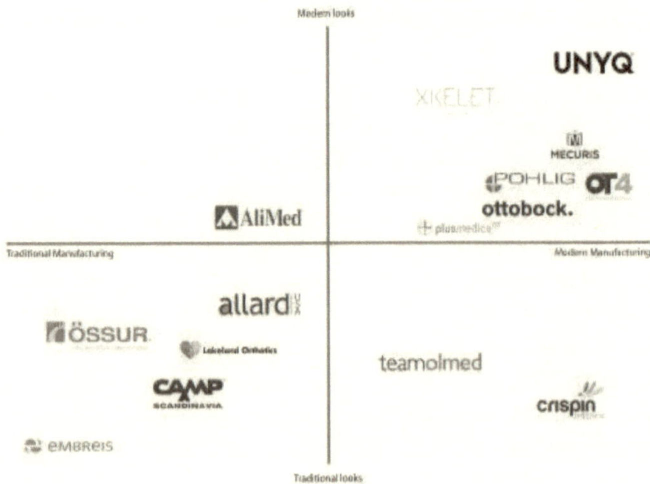

Figure 17, Perceptual map including a selection of relevant competitors

Through defining the qualities of the competitors' products on two scales, one can see where there is a potential gap in the market. It is important to note that the goal of this project is to create a design that is serving both more aesthetic looks but also an innovative application of modern manufacturing methods that involve ADM, meaning, it is desirable to compete with brands that reside in the **upper right quadrant** of the map.

In this graph, brands and companies included are not only producers of AFOs but sometimes also orthoses in general. Especially the ones in the relevant quadrant that use ADM methods are rare and it is even rarer to find examples of AFOs made by them. The few examples found are using simple perforations which solve problems mentioned in the user needs, but none of them has answered the aesthetic need in AFOs.

The only competitor brand that puts an emphasis on style and aesthetics is UNYQ, a company that is not even a direct competitor since they make prostheses, not orthoses. However, their approach is both technically and aesthetically the closest to a solution of this project's task, but since they are not a direct competitor, it leaves the top of the map empty, representing a big market opportunity for technically advanced AFOs that also sell their aesthetic value.

In general, it is confirmed that the AFO market is still mainly using inefficient methods and optimization is desired by both manufacturers and practitioners. There were many other brands worldwide that would have fit into the lower two quadrants but finding competitors for the upper right quadrant was a more demanding, therefore, almost all relevant companies should be represented in that map.

4.2.3 Function Analysis

By doing a function analysis, the most important functions were listed to make sure they will remain in the new design. By assigning priorities to them in main function (MF), necessary function (NF) and desired function (DF), it is easy to track in which hierarchy they need to be implemented. Fig. 18 below shows the functional requirements that the new design must fulfil.

Figure 18, Function Analysis

#	Type	Verb	Noun	Description
1	MF	Support	Ankle	
2	NF	Stabilize	Stance Phase	
3	NF	Minimize	Abnormal Alignment	
4	NF	Minimize	Energy cost of walking	
5	NF	Ensure	Safe position for initial contact	
6	NF	Avoid	Skin damage	Even when worn for long periods
7	NF	Facilitate	Fitting	Donning & Doffing & Inside shoes
8	NF	Maximize	Durability	
9	DF	Maximize	Mobility	
10	DF	Offer	Cosmesis	A patient's need to fit in with peers
11	DF	Reduce	Costs	Compared to current cost efficiency
12	DF	Reduce	Manufacturing Time	Compared to current time efficiency
13	DF	Reduce	Materials	Cast & Product
14	DF	Facilitate	Tailoring	Measuring & 'final fitting'

The function analysis can serve as a checklist to inspect whether the final design concept fulfils all requirements or not. Not all desired functions/DF must be implemented but they would add a great value and selling points to the design concept, so it stands out from the currently available options.

4.2.4 Target Audience

AFO users can be any person regardless of their age or gender, it is a medical device that helps to rehabilitate from accidents, which can happen to anyone, or to help ease disabilities that anyone can be born with or gain to have later in life.

What is more, is that this project wants to offer customization and self-expression. In the end, peoples' tastes differ greatly, there will be no single solution that fits all. Therefore, it has been decided together with the client and project supervisors to create a catalogue of options on a new AFO shape that serves as a blank canvas, rather than one final design only. The design variations will focus on the following target groups:

- **Children**, infants and up to 12-year-old users, male & female variations
- **Teenagers and adults**, male & female variations
- **Elderly people**, from 60 onwards, male & female variations

4.2.5   Design language

There is no brand to work with for this project that defines a set of terms specific for their design identity, so it will be more difficult to grasp what the large audience consents as 'aesthetically pleasing' or at least 'more aesthetic' compared to the current AFO designs. One approach is to define the design language of the currently available designs since the new concept will inevitably be compared to the status quo. Putting the factor of customization aside first, it has been decided in accordance with the project supervisor that there will be one improved AFO shape that makes best use of the new manufacturing method and the new material properties. That new form can be used as a blank canvas for any personalization ideas. The client has expressed that the concept should focus on a solid AFO or SAFO first since any features, especially customization features, can easily be translated to other AFO types if they work well with the SAFO.

Therefore, a design language for **current AFOs** and the new SAFO concept has been defined using mood boards. A visual and verbal representation for traditional AFOs produced and offered by competitors are shown below in Fig. 19.

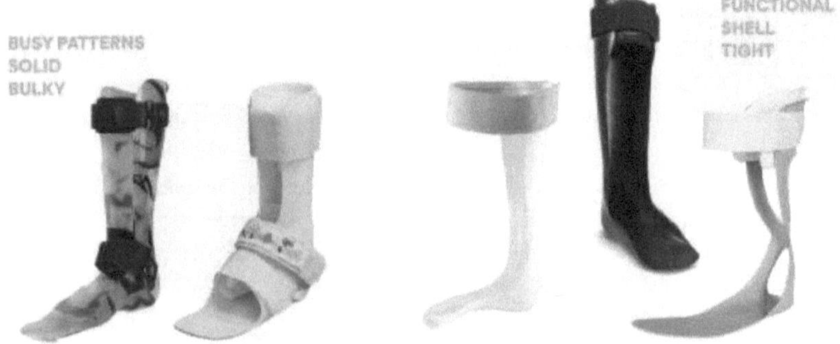

Figure 19, Design language of currently sold solid AFOs

By looking at the currently offered AFOs, it is possible to define what the new design should not be. The design language of traditional AFOs can be explained by words such as bulky, static, solid, tight, improvised, functional, enclosed, and inconsistent.

On the contrary, the currently **existing 3D-printed orthoses** have not explored all aesthetic limits yet, the few examples found are using simple perforations which solve some problems mentioned in the user needs like air permeation, but otherwise look almost the same as the traditionally manufactured AFOs. None of them have responded to the aesthetic needs of AFO users, but some examples are shown below in Fig. 20.

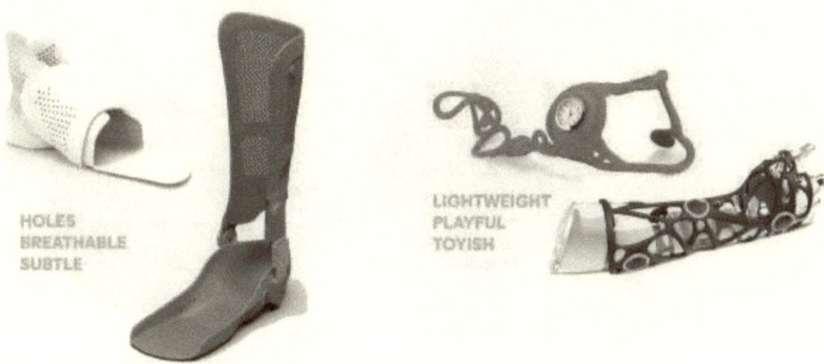

Figure 20, Perforations in wrist orthoses (left) and 3D printed wirst orthoses for children (right)

The perforations alone, that some competitors use, are too simple and neutral but functional (see left side of Fig. 20), while the playful design language of other competitors (see right side of Fig. 20) did not seem appropriate for age groups other than the children while this project aims at offering a solution for all ages. Most companies do not seem to work with an industrial designer or a design department, which is why they release a random design, which can look more unique than traditional AFOs, but they are not necessarily more aesthetically pleasing to the big crowd or they mostly appeal to children. Attributes to describe them would be playful, toy-like, friendly, and bouncy, but it does not cover design moods such as elegant, subtle, fashionable, stylish, modern, or adult.

With the design mood of the current traditionally produced AFOs and available 3D-printed orthoses defined, the **new form language** to be aimed at can be described using words like open forms, see-through, breathable, light, sleek, subtle, elegant, intentional, visible function, intuitiveness, pleasing, and fun.

Furthermore, an image board for the new form language has been created and is shown below in Fig. 21, compiling products where straps are smartly integrated, transitions between functional parts are fluid, perforations are used in a functional but also decorative way to match the product's form. Any moveable parts are accented accordingly so function is well communicated.

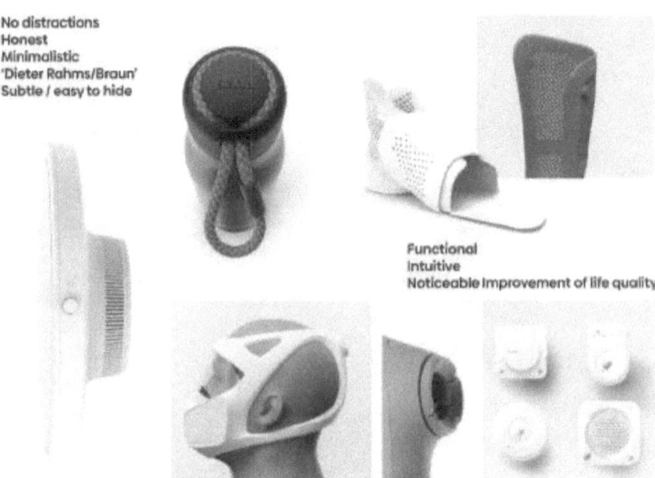

Figure 21, intended new design language

To apply any patterns or perforations in a smart and elegant way, another source of inspiration was parametric design which uses algorithms in computer software to apply patterns to a surface. Those patterns would be aligned to a surface so they can react dynamically and adjust themselves to the surface in case it shrinks or expands, which would come in handy for AFOs that will have many different sizes depending on the user. A mood board for the parametric patterns was created to visualize possibilities that can be expressed with the patterns or textures and is presented in Fig. 22 below.

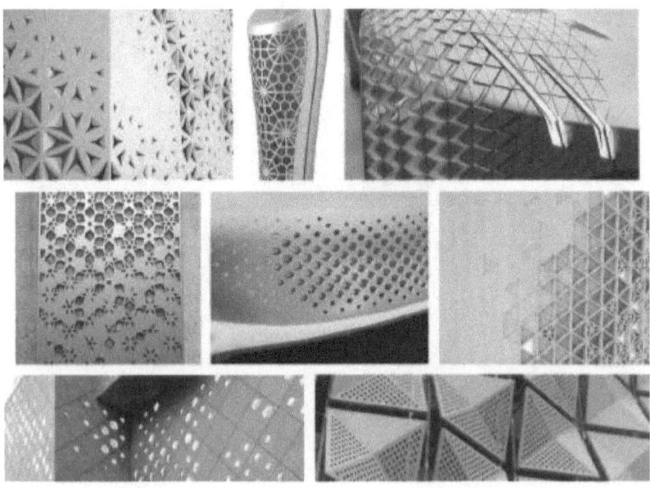

Figure 22, Parametric design and parametric pattern ideas for perforations

4.2.6   Printing technology

By utilizing the newest ADM techniques, new potential can be unlocked to give AFOs a new appearance and functions. To make the best out of the technical potential as well as to stick to the limits, the specific printing technology needs to be defined, so that the designed concept and its process will be as feasible as possible. Through literature review, internet research and competitor analysis, the most popular and most reliable method to print an orthosis with was **SLS: Selective Laser Sintering**. [35] It cannot be proven in this project that SLS is the only correct printing method for this purpose, but it is the method that is currently most prominent to make orthotic devices, therefore the competitor's standard, which is SLS, or to be more precise Jet Fusion by HP will be set as the technical standard in this project to create 3D prints [40].

Jet Fusion is an SLS method invented by HP, who are also producing 3D-printers for other technical industries and already are suppliers for some private orthopaedic clinics that belong to the competitors, providing them with printers and 3D scanning equipment, e.g. Crispin Orthotics. By implementing Jet Fusion by HP, the new concept can be designed with the knowledge that it is possible to control the colour and translucence for aesthetics, stiffness or elasticity for functionality, surface texture to manipulate wear or friction, or even the material printed in a certain spot. All of that can be influenced in steps of the smallest printable unit called voxel, which has a size smaller than 1mm [41].

4.2.7   Product material

During printing technology research, the sources also revealed the most popular and frequent materials used for printing orthotic devices. The most preferred materials were **nylon materials**, more specifically **PA11 or PA12**. When looking up Jet Fusion printers and techniques on the HP website, case studies by Crispin orthotics and OT4 can be found, revealing other applications of Jet Fusion in the medical or wearables sector. The case study reveals more downsides of the current production methods which are that the use of carbon fibre is not as environmentally friendly, difficult to adjust and finish and an expensive to produce material. Crispin has reduced the costs by 50% compared to the use of carbon fibre for a dynamic AFO and now uses a highly reusable PA12 nylon material, enabling them to change the design to be more lightweight and slimmer for better cosmesis [41]. The positive effects of the highly relevant case study and other publications [42] lead to the decision that PA12 or PA11, which is also bio-degradable, should be the material to be printed with in this project.

4.3   IDEATE

The ideation phase is an iterative process, where several methods will be applied repeatedly and in cycles to generate a final concept. In this project, the ideation process is split up into two phases, with a break in between that is the Mid-Presentation, held on the 18th March 2020. The paragraphs within 4.3 IDEATE will explain the process of creating ideas in order to present concepts for feedback at the Mid-Presentation first, then moves on to a more thorough and deeper ideation phase that develops the final design that already implements all previously received feedback from the Mid-Presentation. The ideation process from the second phase ultimately leads up to a finished concept, ready to be turned into a prototype in the next phase, so it can be presented at the Final presentation on May 13th, 2020. The following sub chapters are written in a chronological order, following both ideation phases.

PHASE I

4.3.1   Anatomy Study & Silhouette Ideation

The first practice for warming up before moving to sketching AFOs was to conduct a drawing study with legs and feet to get to know the anatomy. In the next step, AFOs were drawn around the leg to get a feeling for the proportion of the device with the leg. While the pen is moving in the hand, the form of legs and the specifics of an AFO are processed and remembered. Sketching was done with both traditional and digital media like pencil, paper, but also a drawing tablet. At some point in March it became difficult to access the campus for scanning due to recommendations to avoid public spaces to slow down the spread of COVID19, so sketching was done completely digitally. Before moving on to sketching detailed AFOs, the desired forms and how they would fit around a lower leg were expressed by painting one solid area like a silhouette only over the legs. A few examples are shown below in Fig. 23.

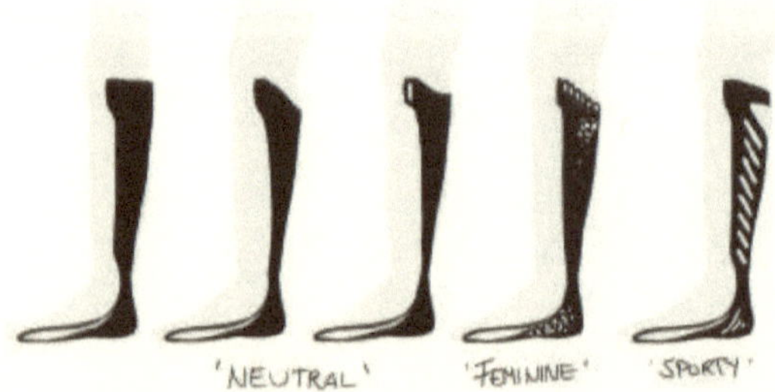

Figure 23, Silhouette drawings over sketches of legs

4.3.2   Concept Sketch Ideation I

The first round of ideation is rather loose and serves the purpose of expressing first ideas that solve the user needs mentioned before without trying too much to cover up all age groups. The main purpose was to create three different concepts, so they can be shown and discussed at the Mid-Presentation. All concepts were presented with rough colouring and shadowing to enhance the realistic impression of them. For the presentation, not only the sketches themselves but also detail shots, process sketches, key words, and images to describe where the inspiration came from where added to each concept. The first three ideas presented were as follows in Fig. 24, Fig. 25, and Fig. 26:

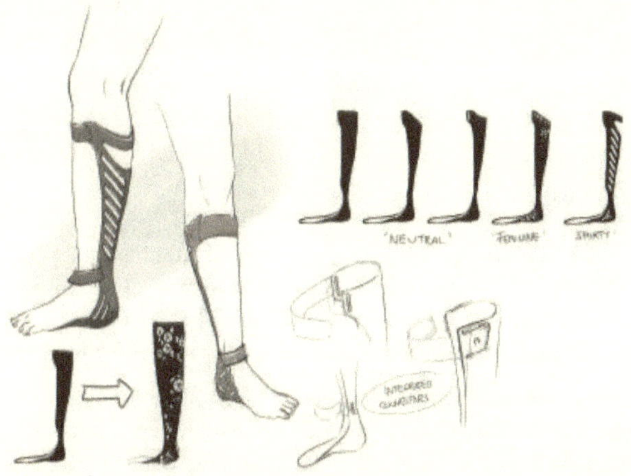

Figure 24, CONCEPT 1: parametric design applied to an existing shape with alterations

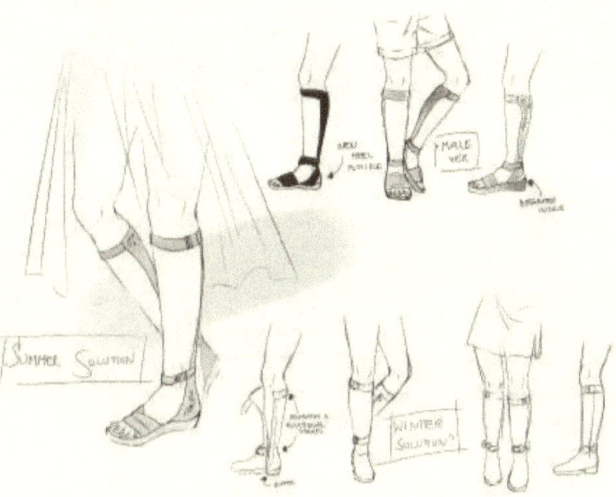

Figure 25, CONCEPT 2: integrated shoes/soles and fashionable straps

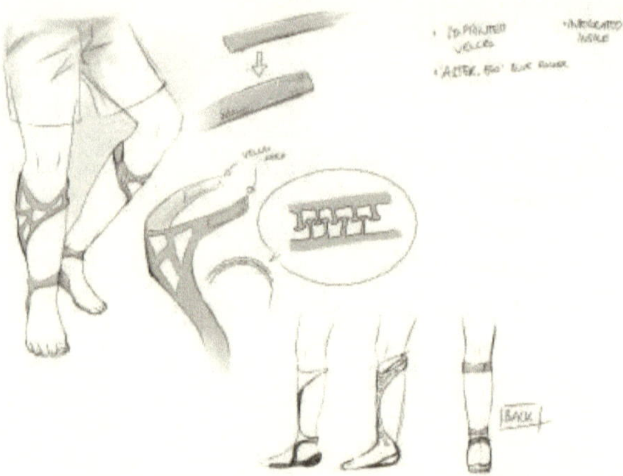

Figure 26, CONCEPT 3: fluid shape for Blue Rockers kind of AFO with integrated printed Velcro strap

4.3.3   Feedback Session: Mid-Presentation

Presenting concepts and receiving feedbacks is a means of testing but can also spark new ideas when others can express their opinions and inspirations. All participants from the Mid-Presentation were able to give feedback and input. The most relevant feedbacks were given shortly after the presentation by the client and the supervisor of the project.

The client Nerrolyn expressed that she liked the 'sporty' pattern of Concept 1 and prefers its simplicity. The printed soles from Concept 2 were experimental but would mean that users cannot change their pair of shoes anymore when they want to wear their AFO. They would be stuck with the same design of the shoes until they order a new device. In terms of what people would rather want to change, it would be more practical to be the pair of shoes rather than the whole AFO. A wide variety of AFOs with soles would be needed as opposed to a variety of already existing shoes instead and designing that variety takes a lot more effort than letting the patient pick shoes to their taste from other retail options. Furthermore, Concept 3 was not very relevant as it is based on a dynamic AFO. In her opinion, Concept 1 is the most feasible to move on with, since a design that can be made for a solid AFO it is very likely to be applicable to the surface of a dynamic AFO as well. However, Concept 3 was a good representation of how a pattern could look like on a DAFO shape.

Further feedback received from the project supervisor was to continue with **Concept 1** as well if the client has already expressed their desire like that, and advised to make the final product one optimized SAFO and to produce an additional catalogue of patterns that caters to every target group. The patterns can be implemented on the AFO surface as cut-out holes or perforations. It was also recommended to move on to ideating with 3D modelling as soon as possible.

PHASE II

4.3.4   Brainstorming

After determining that it will be one basic SAFO design with various patterns, to get into the new theme and to gather ideas for patterns, brainstorming for words to describe the tastes of the different target groups was a simple method to empathize with the new task and to verbalize the themes to go for when creating the designs. This session was based on experience and assumptions from common impressions of the fashion observed in daily life. The reason why this was done before further market research, was so that the ideas were not tainted with things recently perceived and stored in short term memory but to also pull out ideas from long term memory that might be suppressed or hidden under recent visual memories. The words were written down and connected in mid maps using simple tools like a piece of paper and a pen, but then are sorted digitally and are shown below in Fig. 27, Fig. 28, and Fig. 29.

Figure 27, Words describing patterns and themes for children

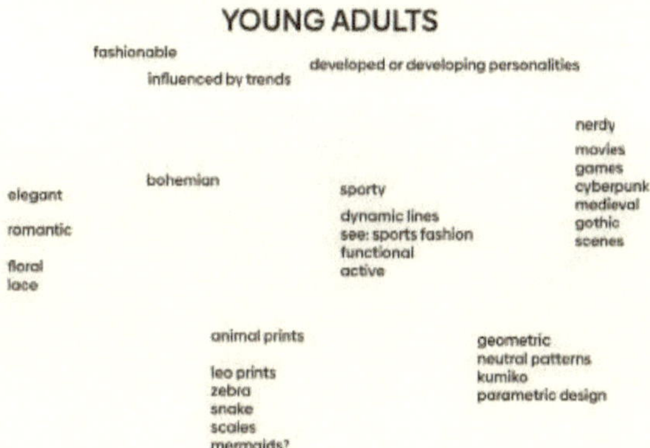

Figure 28, Words describing patterns and themes for teenagers and adults

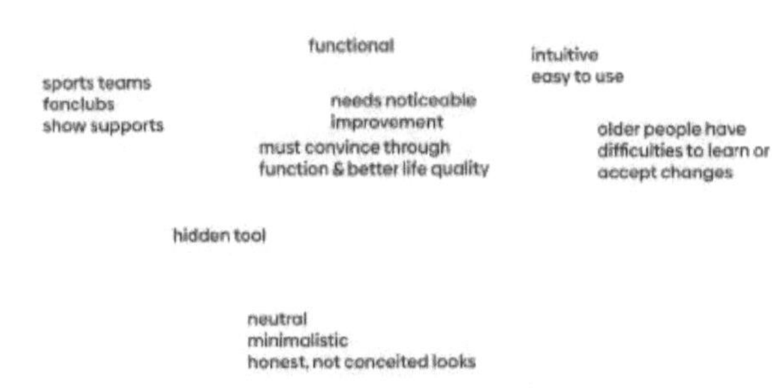

Figure 29, Words describing patterns and themes for elderly patients

4.3.5   Trend Analysis

The next step was to find out which kind of patterns are relevant. It is important to use pattern themes that people like enough to wear on other pieces of clothing. To take a shortcut, current trends are easy to spot by looking at what the fashion industry, that has done their research, is offering. Clothing items with those popular patterns are easy to find by browsing through their online shops. Looking up clothing items online also serves as a trend analysis, since the items presented are from the current Spring or Summer collections, which have ever changing themes every year. The online shops chosen to be investigated are common fashion chains such as H&M, Zara, and Topshop but also the website of the largest fabrics supplier to consumers that is or its Swedish equivalent

After some trend spotting, the decision was made to deliver at least two pattern designs to every target group. From collected reference images online, two mood boards are composed that show stylistic choices commonly seen in clothes for children, seen in Fig. 30, and adults, seen in Fig. 31, using both visual and verbal tools.

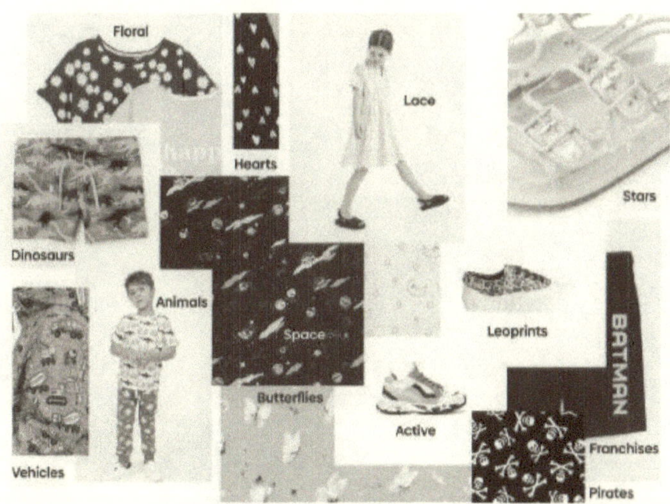

Figure 30, Popular patterns for children

Figure 31, Popular patterns for adults

4.3.6   Stylistic Choices

An idea that was left from Concept 1 in the first ideation phase was to integrate the belt straps into a belt loop hole that is part of the printed AFO. Another image board shown in Fig. 32 below contains various inspirational sources using images from consumer products and fashion items that use straps. Things to consider were the material combinations, how they were cut or knotted together, and how the transitions into the solid and stiff main body worked.

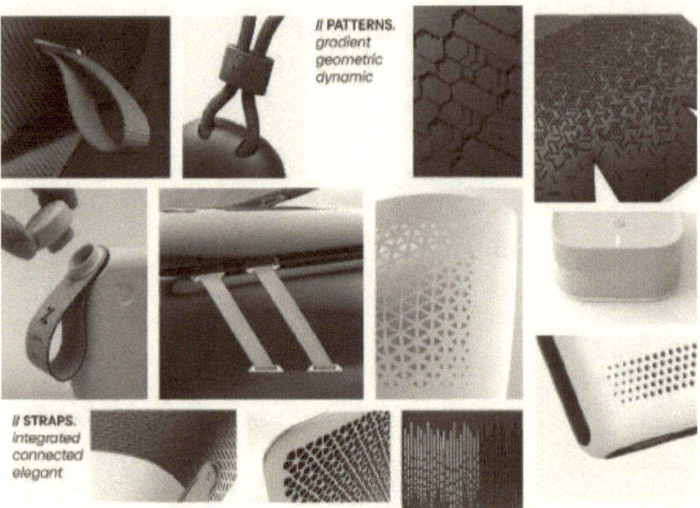

Figure 32, Inspirations for strap attachments (left) and arrangement of patterns (right)

Air permeability, which was an important user need, can be fulfilled with having cut out patterns. Inspired by the indirect competitor UNYQ that spread design patterns across the surface of their protheses [48], patterns can be applied to the surface of an AFO using parametric design. Which means that with the help of computer algorithms, it is possible to apply any pattern to any AFO size or type and through automatic calculations it will spread out accordingly to match every size and any surface boundaries. The right side of Fig. 32 above demonstrates what style of parametric design is aimed for. It has many geometric patterns, mostly made out of some sort of polygons, arranged to look like stars or shapes and are stretched over a surface in a gradation where a shape changes in size or in distance to the next pattern, which serves an interesting and dynamic look rather than simply pasting a pattern that fills up to the borderlines of a surface.

4.3.7　Concept Sketch Ideation II – after feedback session

After deciding to take Concept 1 from Phase I as the starting point for Phase II and designing the basic shell of an SAFO with one strap on top, the next step was to ideate for the design of that basic shell. When analysing which part could be changed, a schematic sketch shown in Fig. 33 below helped in marking which areas can be changed by how much, but also which ones must remain the way they are.

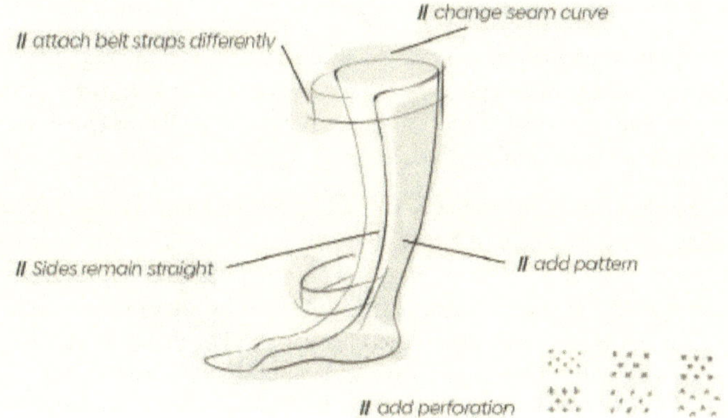

Figure 33, Schematic overview sketch of the design elements of a SAFO

When searching for fashion trends for AFOs, the result was that an AFO needs to fit under a pair of pants that can be relatively tight, leaving little space for another layer embracing the leg, therefore proper fitting needs to be considered and the thickness of the solid AFO should be kept minimal [4].

Furthermore, the general design language mood board that was shown in 4.2.5 'Design Language' is still valid for sketching in Phase II. A selection of sketches that were relevant for the final design are shown below in Fig. 34.

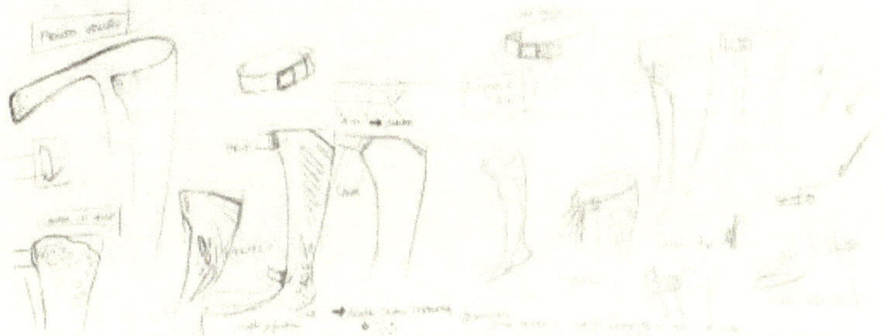

Figure 34, Sketches from Phase II, focusing on curvature and belt loop design

4.3.8   Graphic design: digital sketch ideation

Another part to fulfil for the final design concept was setting up a selection of patterns for the AFO surface, which can be viewed like a catalogue. For each target group, an image board was created in the trend analysis as previously shown in 4.3.5 'Trend Analysis'. To find suiting images for the mood boards, the internet research focused on existing multicolour patterns used in other AFOs, fashion items and fabric patterns. The mood boards serve as a source of inspiration when creating graphics for patterns.

Because the pattern will be expressed with holes, it is important to keep in mind that only closed silhouettes can be used to make shapes, so all patterns created must work as a black and white graphic and each cut-out must not have details inside because that would require the material to float which is impossible.

All patterns were drawn digitally in the software called **Clip Studio Paint** on a tool called vector layers that allow loss free scaling of lines that have been drawn. Despite being called vector layer, the graphics are not saved in a standard vector format and are treated rather like pixel images, meaning that upscaling outside of the software would not be loss free in quality. However, working with a digital pencil and eraser like that is more intuitive, and design changes are quicker to implement than by drawing with vector programs like Adobe Illustrator. When a pattern is finished, it can be exported as a high resolution and high contrast PNG image, which is then imported into another software called **Inkscape**. It is a free software that has a reliable plug-in to convert high resolution pixel images to vector lines, which saves a tremendous amount of time compared to tracing patterns manually. The plug-in is especially handy when free forms with many curves need to be converted like animal or floral patterns. The selection of all 21 patterns that were created with all themes and target groups is shown below in Fig. 35 and all patterns on a larger scale are to be found in the appendix.

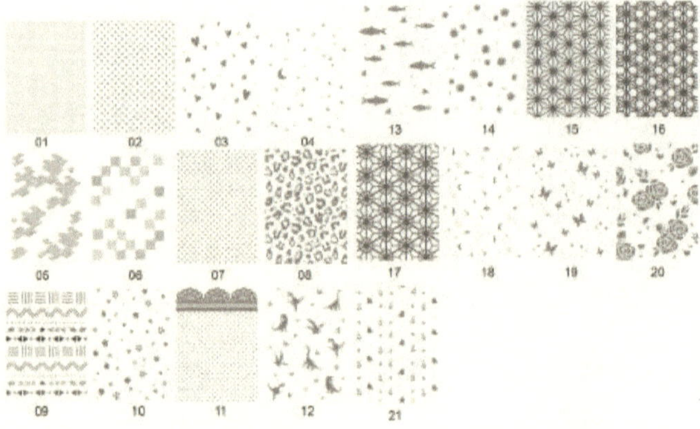

Figure 35, Pattern catalogue with all 21 designs for various target groups

4.3.9   Kansei Study & Online Questionnaire

To make the most out of the remaining time available for the project, it would be best to implement a relevant pattern that fits the design intention as much as possible for the 3D model, meaning that the physical prototype will reflect typical patterns that represent each target group best.

To find out which one of the 21 pattern designs were fitting representatives, all of them were uploaded online for a questionnaire using Google Forms. Participants were asked to rate the patterns on scales that quantify their perception of its gender association, age group belonging, and personal emotional value towards each pattern. Additionally, they were asked to name some of their personal favourites. An example of the interrogation is shown below in Fig. 36 using one pattern as the example.

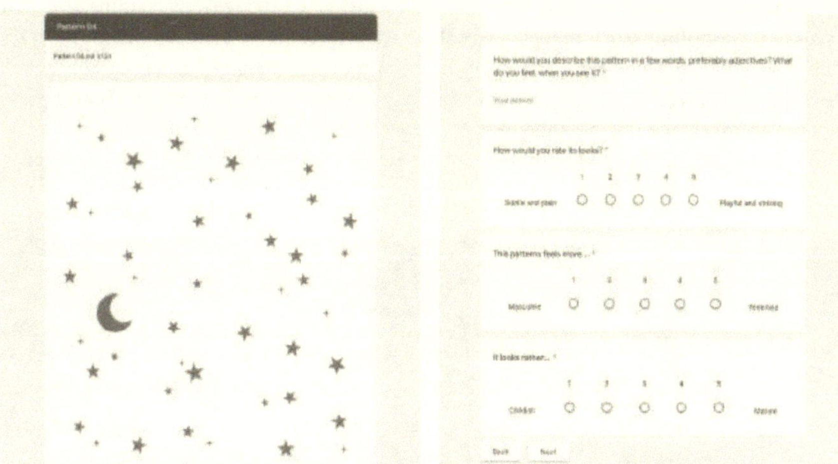

Figure 36, Questionnaire example that is applied in the same way to every pattern

From the overall results gathered within a week only due to time pressure, the most popular patterns from the questionnaire were picked, which are shown below in Fig. 37. The questionnaire was left running until the end of May to gather more data which will be relevant in the TEST phase which is described later in the report.

Figure 37, Patterns most associated with male users (left), female users (middle) and children (right)

4.3.10 Ideating through CAD

Further changes in designs were explored through 3D modelling in Rhinoceros 3D, e.g. the thickness, curvature, angles, and patterns were built in an experimental manner. Hand drawn sketches were made in parallel to quickly visualize and note down ideas on paper first before they are implemented into the 3D model.

When modelling in Rhinoceros, it is important to remember that some geometries cannot be reverted back once they are trimmed or filleted, so when trying out variations, it is best to make copies of the latest model, move them to different layers, and try out ideas on the copies without damaging the surfaces of the latest model version. An example of how working with several copies of a model works is shown below in Fig. 38. The purple model (far left) is the raw base that was copies several times to explore different trimmings, cut-outs, and perforations until the final design (far right) has been achieved.

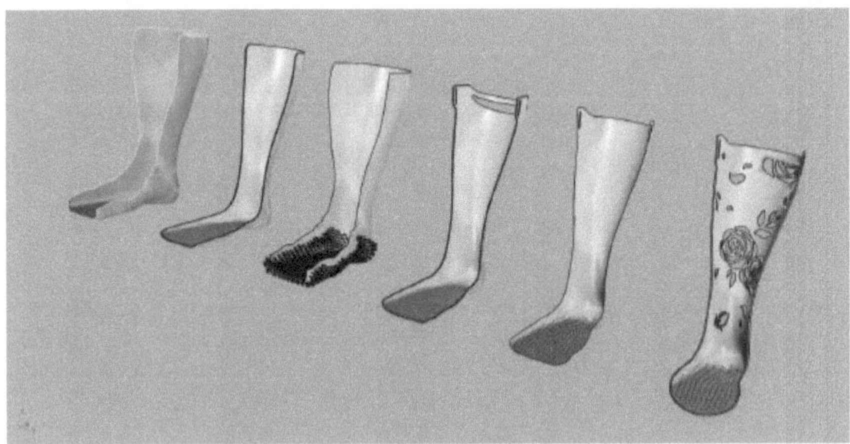

Figure 38, Ideation with copies of CAD models in Rhinoceros 3D

4.3.11 Weekly Online Feedback Sessions

Shortly before the Mid-Presentation, project supervision was moved to an online environment, using Microsoft Teams. Weekly calls with the project supervisor to check in with the results were helpful since they were a change to ask for quick feedback to what has been achieved the past days. They lasted for 15-30 minutes each time and were helpful as a motivational tool and to ask for feedback or technical help with 3D modelling or report writing, steering the project in the right direction.

4.4 PROTOTYPE

4.4.1 3D Scanning

The first step to make the prototype or rather appearance model was to take measurements of a leg which was taken from the leg cast that was a by-product of the role playing done at the beginning of this project. It simulates the planned process of the actual production method that is to use a 3D scanner but for this project utilizing the cast from a traditional procedure instead of scanning from the live tissue. The scan was made using a 3D scanning equipment called **GoISCAN50** by CREAFORM, which is a handheld scanner that scans with shooting light patterns that are then bent by the specimen. From those curvatures, a 3D mesh out of polygons is created, therefore the scanner only works when connected to a computer with the right software running. During the process, a teacher who also happens to be one of the interviewees was assisting. To differentiate the ground plane from the leg, some locator stickers are placed on the desk, where the cast was resting on, to mark its position as shown below in Fig. 39.

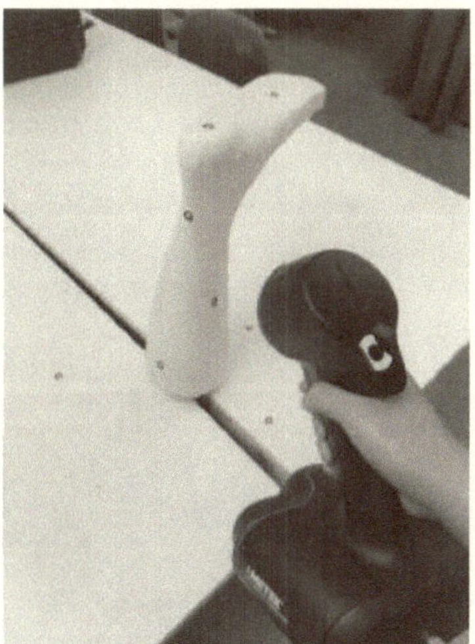

Figure 39, In progress of 3D scanning the leg cast

The scanned file was later imported into **Meshmixer** by Autodesk where holes and uneven spots were cleaned up to provide as clean as possible curves. Once the model looked smooth enough to work with, it was converted from a surface model to a solid one and exported as an STL file that can be imported into the desired 3D modelling software. Fig. 40 below shows the scanned leg before and after fixing the mesh.

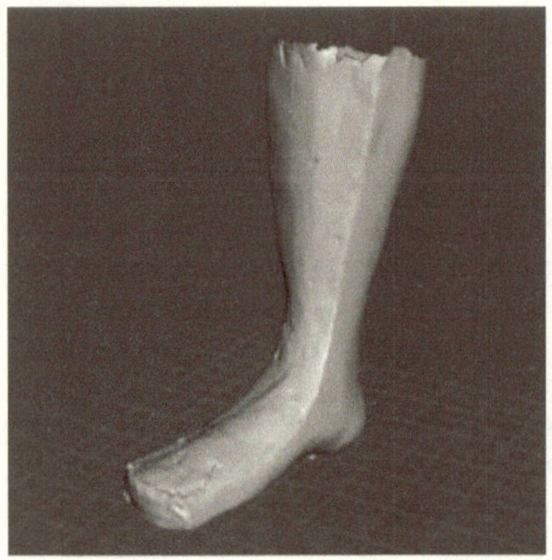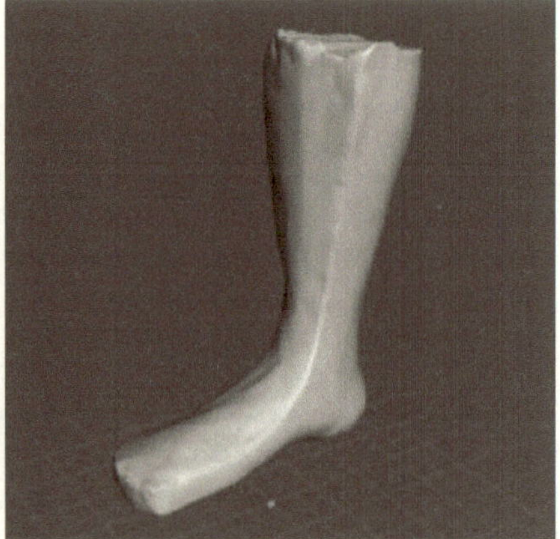

Figure 40, Scanned leg 3d model before (left) and after (right) cleaning up holes and restoring edges

4.4.2   CAD / 3D Modelling

Rhinoceros 3D was used to model the more organic shape of an AFO, using the cleaned 3D scanned model as a reference. By using a snipping plane, the model was cut with a moveable planar surface that reveals the outlines and the cross-section surface of the model as shown below in Fig. 41. Those borders visualized with a black, thicker line are then traced to make 2D curves. Sometimes even 3D curves were converted from two 2D curves of different perspectives, when necessary for transitional parts like the heel.

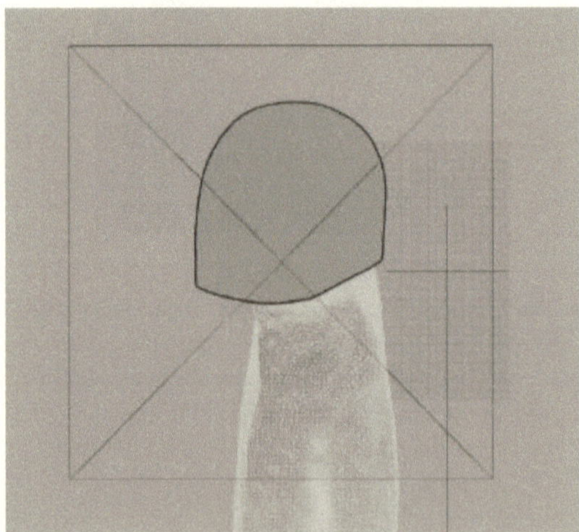

Figure 41, Cross-section from above shown by using a clipping plane

From those 2D and 3D curves, the organic leg surface was built using a loft, which was later trimmed to the desired dimensions for the border design. To create a solid body, an offset in the desired thickness of 2.4mm was made. The lofted surface and the 2D and 3D curves are shown below in Fig. 42.

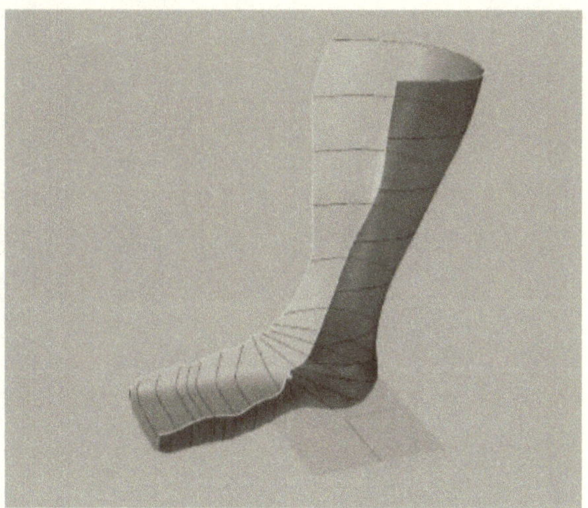

Figure 42, Lofted surface of the posterior leg shape using a network of 2D of 3D curves

Lastly the desired pattern was integrated using the commands called 'create UV curve' and 'flow along surface', which flatten the loft surface so lines can be drawn on it without distortion. Further, curves and extrusions are projected back onto the original leg shape, wrapping them around the surface position that was picked from the flattened image. An example showing the flattened surface with aligned patterns and the projected extrusions is presented below in Fig. 43. It shows the curves (left), extrusions on the flattened UV curve surface (middle), and the final AFO with projected and trimmed patterns (right).

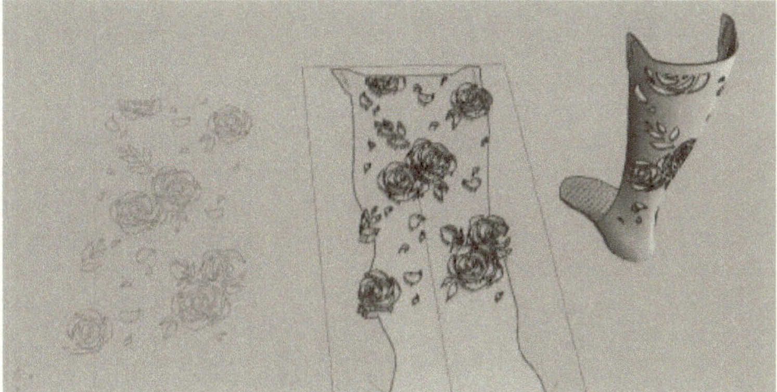

Figure 43, 3D modelling process of the decorative pattern

Unwanted holes were trimmed or deleted, and sharp corners were filleted or chamfered. Before filleting as the last step, it is important to close and combine all surfaces to make a solid body. Whether that was successful or not can be checked within Rhinoceros; when the result says that the polysurface is valid, it means that all surfaces are without errors and connected to each other.

4.4.3 Rapid Prototyping: Mock-Ups

The first physical models were test prints made on campus using Makerbot Replicator+ printers with the goal to validate the sizes, ensuring proper fitting, and design choices before making the final prototype. All printed mock-ups are shown below in Fig. 44.

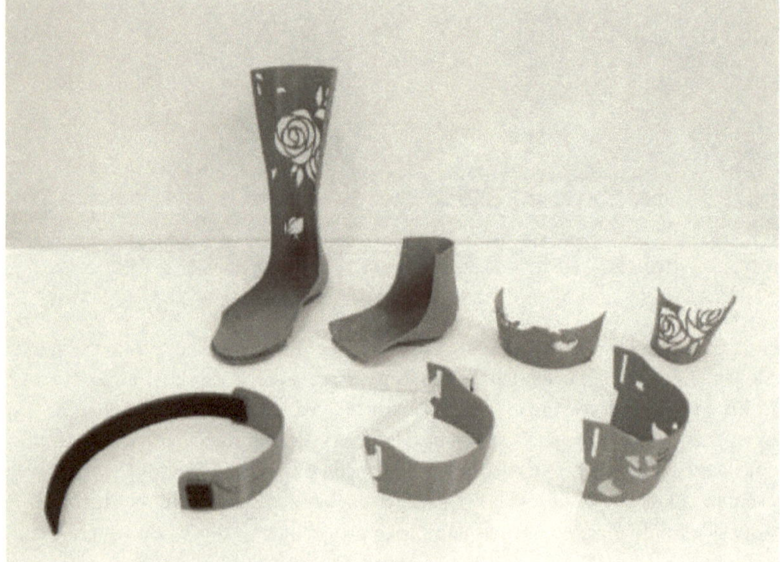

Figure 44, Printed mock-ups using an FDM method in Replicator+ printers from university

After it was confirmed that the pieces fit, more mock-ups including the first strap hole designs and patterns were printed to check if the perforations are visible and feasible to have for the appearance model. Additionally, different Velcro straps were added to the prints for testing the strap and strap holes function as presented in Fig. 44 above as well. Rapid prototyping is therefore a tool to physically test designs and change them again after evaluation. It is partly for ideation but also for testing, which will be further explained in the TEST phase.

4.4.4   Appearance Model

For making the appearance model, the same printing method as for the mock-ups was used but the surface quality enhanced further with additional steps. The initial idea was to print the AFO model in one piece using the special printer Z18 that has a bigger and higher building plate but works the same way as the Replicator+ that was printing the mock-ups. However, the printer ran into a hardware error when a quarter was left to print and it could not be solved without replacing components, thus, the Z18 would not be able to print another prototype in time for the Final Presentation. The failed first print attempt presented below in Fig. 45 was kept as a backup model since it could be partly used up to the point where the error occurred in case alternative solutions would not work.

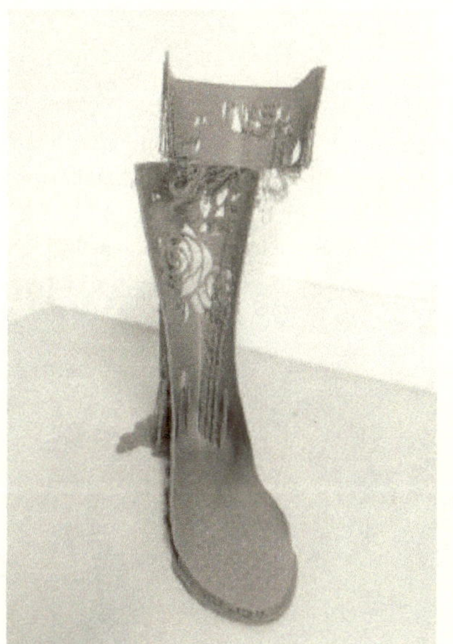
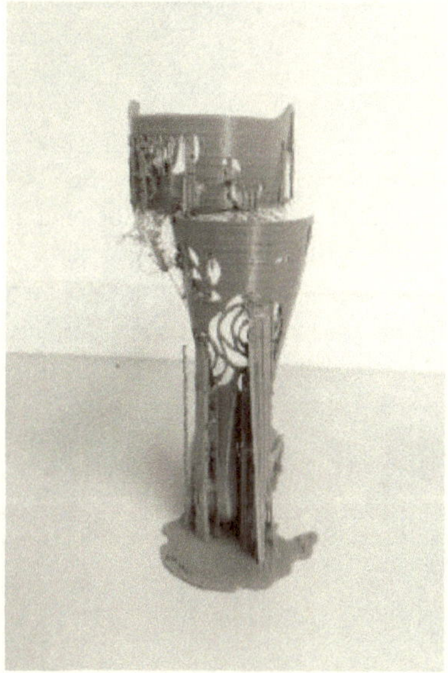

Figure 45, Failed prototype attempted to be printed in one single piece, resulting in a huge layer shift

The alternative for making the appearance model then was using the Replicator+ printers to produce the AFO in three split parts that are joined later. Splitting was necessary since the building height in a Replicator+ machine did not enable printing the AFO in one or even two pieces. The splits were made horizontally at intersecting planes that would impact the implemented rose pattern the least, so joining the pieces and blending the transitions would be simplified. Fig. 46 below shows all three printed parts used to make the actual model before any post-processing, but with all support material removed as far as possible without sanding.

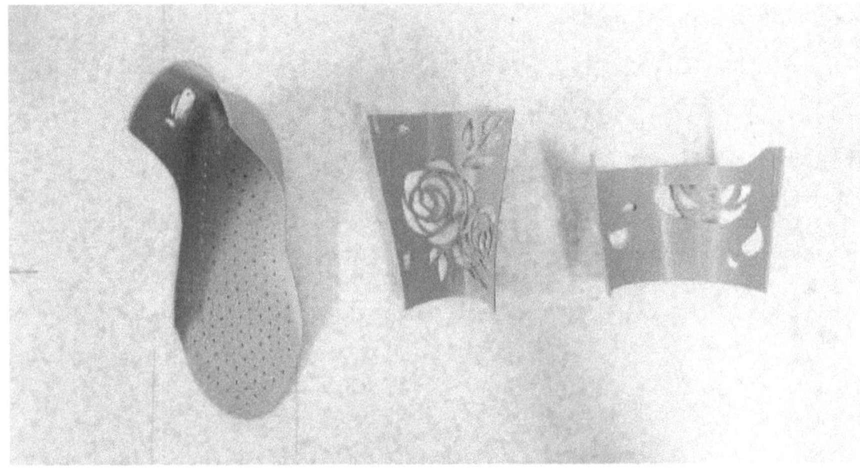

Figure 46, All three printed parts for the appearance model in their raw state

When the print was finished, sanding and cleaning up the surface with only the PLA material from the filament was the first step in getting the surface smooth like the example in Fig. 47 on the left is showing. Further steps were filling up gaps with plastic filler material, then joining all three parts into one as seen below in Fig. 47 on the right.

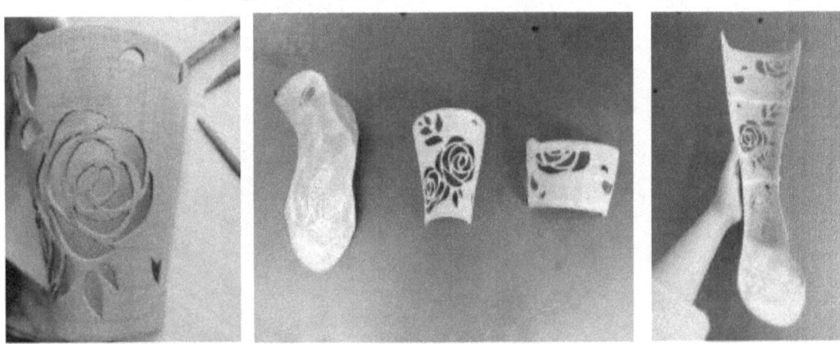

Figure 47, Post Processing 1: sanding PLA (left), filling up gaps with plastic filler (middle), and joining separate parts (right)

The final steps were adding spray filler and sanding it down again several times to give the surface a smooth finish that is ready to paint. The process is roughly shown in Fig. 48 below.

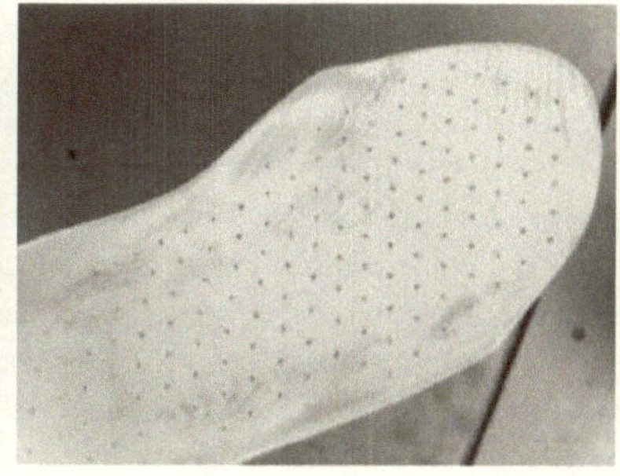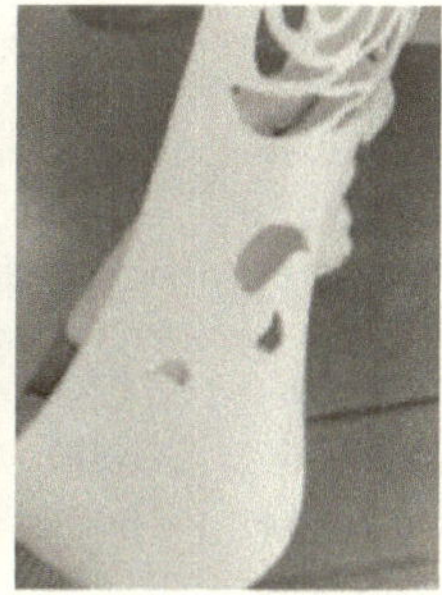

Figure 48, Post Processing 2: Spray Filler and sanding, more spray filler until all gaps are even

The last steps before finishing were to add layers of spray paint in the desired colour on top of the layers of smooth filler. The paint layers were applied five times to achieve the desired matte surface texture and to make sure all surfaces of the relatively big AFO model were covered in black paint and no bright spots from the spray filler with light grey colour were left. The last steps are presented below in Fig. 49. To make the appearance model become wearable for photos and testing, a Velcro strap was sewn on through the loops by hand, the result is shown in the next chapter.

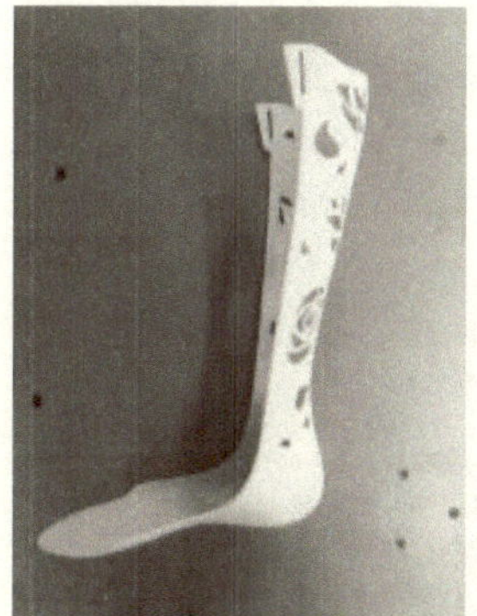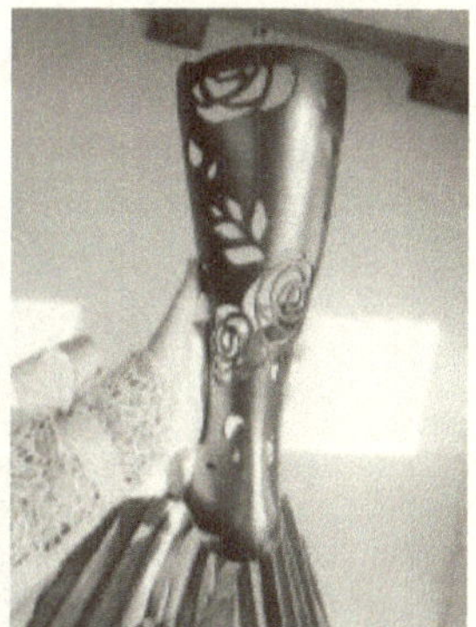

Figure 49, Post Processing 3: Surface ready to be painted (left), after two layers of paint (middle) & after 5 layers of paint (right)

4.4.5 Simulation Rendering

Keyshot that is a software for photorealistic rendering with 3D models and was used for colour variation tests and surface quality check. Evaluating smooth surface transitions and fillets or detecting sharp corners, holes, or overlaps are crucial to conduct so one can make sure that the 3D model has been constructed without errors, which can impact the quality of the 3D print. When in doubt what colour variations to pick, Keyshot can also simulate colours realistically which is a quick and useful means for making design choices without wasting material and time for more mock-ups.

When using Keyshot, 3D files imported should be STEP files to ensure that the surfaces will be smooth and not rough from the triangle structure that is the nature of some file types like STL. The advantages of checking the model in Keyshot over using the respective 3D modelling software only are the real time rendering engine and much smoother appearing surfaces by default. The previews generated in 3D modelling software are often coarse and rough, or surfaces disappear when zooming in, making it difficult to track for corrections. A comparison below in Fig. 50 shows the exact same model but with rendered views in Rhinoceros 3D (top) that appears to have ribs or folds and Keyshot (bottom) that expresses how smooth the surface is to the point that it will reflect like a mirror when the paint is chosen to be glossy.

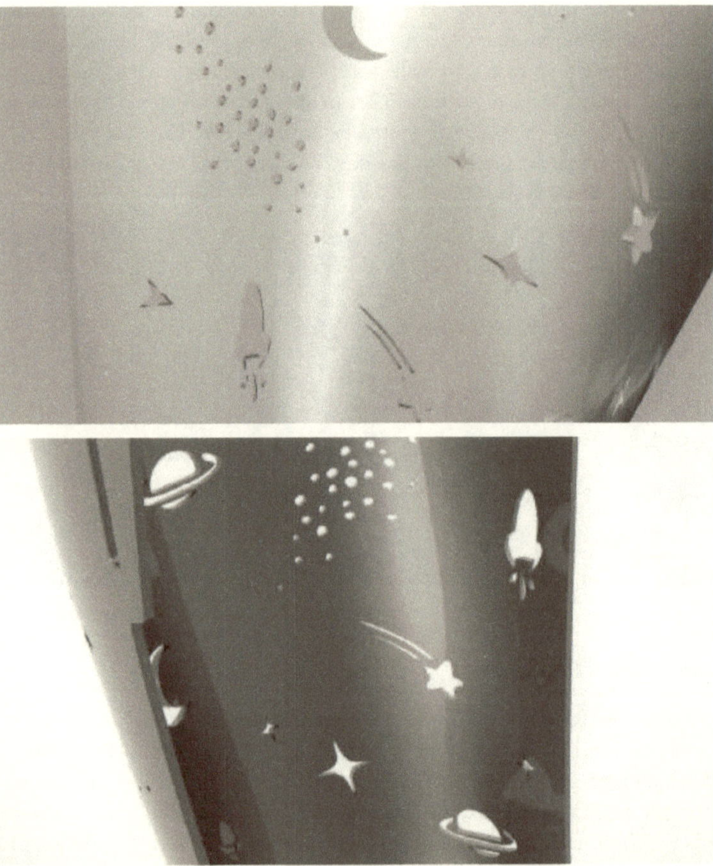

Figure 50, Different surface quality appearances depending on the program used for rendering

4.5   TEST

As mentioned at the very beginning of this report in the disposition chapter, this project does not include any testing with a working prototype to prove its function or testing with end users of the target group. However, there were several other tests conducted to validate design choices, a smooth workflow, to check the quality or find evidence for

4.5.1   Kansei Engineering questionnaire

What was originally planned for a focus group to test patterns and designs had to be moved to an online questionnaire which is the closest that could be done to replace the user test that could not be conducted anymore due to restrictions and regulations concerning meetings with groups of people after the outbreak of COVID19.

The questionnaire for this testing is the same as the one used for making a design decision before prototyping but being left running for three more weeks after the final presentation to gather a bigger number of results. The goal of the remaining testing phase was to determine which patterns ultimately have potential to be part of a permanent pattern collection or which themes are the most interesting to work with.

Fig. 51 below shows the example of an answer sheet to corresponding questions of one pattern (left), submitted by one testing person and their feedback to the whole catalogue (right) which was asked for at the end of the questionnaire. The full extent of the questionnaire and the results by all participants can be found in the Appendices.

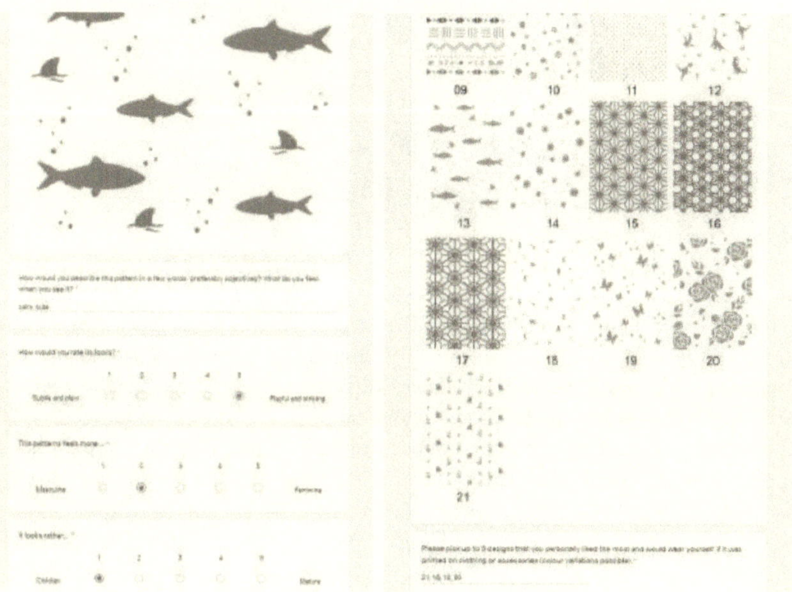

Figure 51, Questionnaire answer sheet example

4.5.2   User testing for fitting & function

The action of printing mock-ups is not only to stimulate the designer's creativity, but also to test function and measurements. Since the dimensions of the AFO are based on the leg dimensions of the design student, the only test person to make sure the fitting was well, was the designer themselves, making this testing more like a case study. To make sure that the scan, that served as a template, was having the correct proportions that fit the user, in this case the designer themselves, critical parts like the heel and a piece of the shell at its highest and widest position were printed from the first CAD model finished. The printed mock-ups were placed around the leg but also on the leg cast to test the fitting as shown below in Fig. 52.

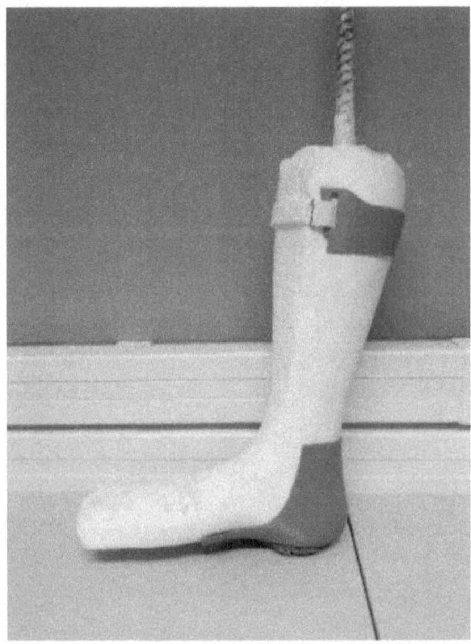

Figure 52, Mock-up tests on the leg cast with 3D prints & Velcro strap

Fortunately, the mock-ups were fitting and even though the AFO felt slightly wider for the user due to weight loss since the time the leg cast was taken, the flexible PLA material and the Velcro strap used to tighten the AFO around the limb were able to adjust the dimensions for a tight and safe fit nonetheless. The mock-ups have also helped finding out what printing settings are best to used and the result is that all supporting constructions are crucial to print the detailed decorative patterns cleanly, so that they do not need to be filled with more material.

4.5.3 Simulation Rendering: colours & textures

As mentioned before in 4.4.5, Keyshot is also used for generating visuals with colour and texture variations. But since any colour is possible in theory, this method was not to pick a definite colour to exclude any others but to better visualize different, possible variations to show that the large colour palette is an advantage. Using digital aids like 3D simulation in Keyshot does not only save money and time but it is also environmentally friendlier than trying out colours with making prototypes and paint.

The exported renderings are also useful for communicating the result during the final presentation and Keyshot as the tool offers many tools for customization such as moving around lights, but also the material library contains useful assets that come as a part of the software. A preview of how the workspace is set up with the material library on the left and the environment and lighting setup on the right is shown below in Fig. 53.

Figure 53, Keyshot workspace layout with the material library (left) and environment and camera set-up (right)

5   Result

5.1   Final Design

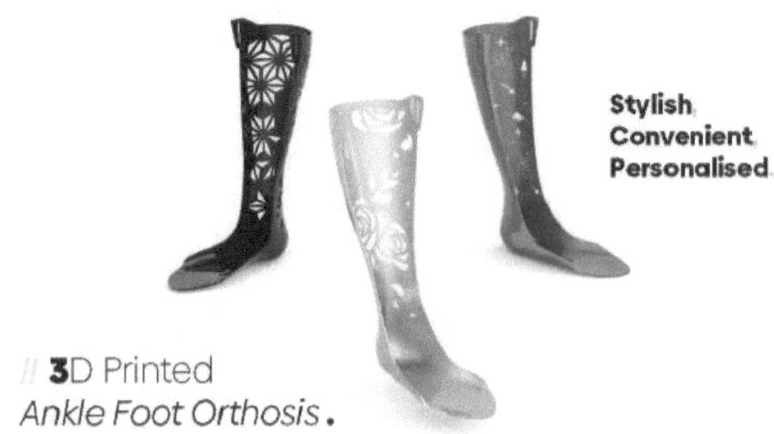

Figure 54, Final design for concept 'AirWalker'

The final result for this re-design project is an ankle foot orthosis concept made from additive manufacturing methods, or also called '3D-Printing', named **AirWalker** that at first glance differs from conventional ones by having striking patterns all over its surface and coming in various colours and textures as shown above in Fig. 54.

It is a concept that prevails over current designs by offering the possibility to have almost any pattern implemented decoratively on the surface of the AFO for a personalized touch and improved aesthetics, no matter whether the patient wants to stand out and make a statement with their orthosis or hide their device more efficiently to avoid social stigmatization. Not only does the patient have a choice over the pattern but also over any possible colour and texture.

Because the patterns are made by leaving out material, the negative spaces create a lightweight appearance while the orthosis becomes breathable at the same time as it becomes more aesthetic product, which is an innovative trait of AirWalker. A smarter connection of straps attached to the integrated belt loops makes the whole product thinner and more elegant as it appears be one whole harmonic unit.

Each AFO design can be created by the orthotist, a modelling specialist, and the patient themselves together, which is a revolutionary approach. With the help of digital tools like 3D modelling with automated algorithms and realistic computer renderings, a user will know almost exactly what they will receive. By using 3D scanning for taking measurements, errors can be avoided, and a lot of time is saved that can instead be spent on the patient and their rehabilitation, benefitting them and the orthotist.

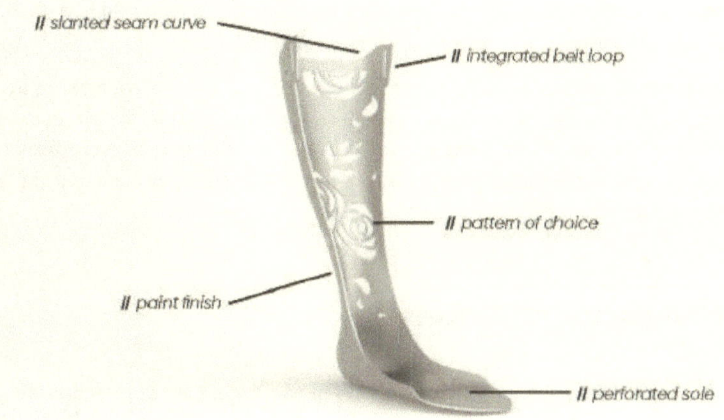

Figure 55, New & improved functions overview

An overview of all new functions and product details are shown above in Fig. 55. New design choices have been implemented from top to the bottom of an ankle foot orthosis. The top starts with a **slanted trimming curve** going along the upper seam, replacing the usual straight cut. A cut in an angle, that points towards to where the straps are, gives the shape a more intentional and wanted look.

By integrating **belt loops** for the fastening straps in the orthosis itself, the Velcro or any kind of strap material is now be better integrated with the shell, making the product as thin as possible so it fits under clothing without sticking out more than necessary. The belt loops are printed along with the main body, skipping one step of attaching external loops and any straps will look more integrated, intentionally placed, and secured instead of instead of their usual out of place position as an extra body.

Any desired pattern can be printed on the shaft if they can be represented in negative spaces. The size of each cut-out may vary as long as they are not too big that they make the shell lose stability from taking off too much material or create sharp corners that might hurt the skin when wearing. **Patterns** can be picked from an existing catalogue or they can be custom made together with a designer. Implementation of existing themes or patterns with a reference or further personalization with names or personally relevant motives are possible as well and all depends on what the end user wants. The big shaft area is almost like a blank canvas for a user's self-expression and imagination.

The perforations from the decorative patterns continue to be part on the **sole**, where significantly smaller shapes were chosen so as much material as possible remains to hold the patient's body weight put against that surface. To withstand that load, it needs more stability and bigger holes with long or sharp edges might lead to injury, pain, or at least unwanted imprint on the skin. Furthermore, the sole part is at an area that is mostly hidden under footwear and therefore needs to focus more on function than aesthetic self-expression opportunities.

In order to prevent other injury and since there was no need for improvement in that aspect, the side trimmings of the new concept need to stay straight in the same manner as the traditional AFOs. Lastly, the final product can be painted or maybe even already printed in the patient's **colour of choice**. Further priming for different textures and other post processing techniques for aesthetic purposes are possible as well.

5.2   Manufacturing Process

SLS or **Selective Laser Sintering** printing with the industry's newest models like **HP MultiJet** was chosen to be the most promising method that enables and encourages exploration of new geometries for AFOs that it is capable of printing in one go. Another significant advantage of using the MultiJet method is its ability to control, colour, translucency, stiffness, and even material type in one print by using different agents to harden the powder contained in the printing bed [40]. If desired the ankle foot orthosis can be primed to give the surface a smoother finish and/or painted in any desired colour.

The selected manufacturing process was not chosen from testing all existing printing methods as part of this project itself but from looking up related peer reviewed articles related to the field of Orthotics and manufacturing and what competitors used including their testimonials and reviews about the methods. It can be assumed that if professional companies that sell orthotic products, developed with medical experts, rely on certain methods, then those methods can be trusted to be effective and reliable to be used in the same context of this project as well. It is therefore the most useful piece of evidence possibly, when other testing methods to gather first-hand date are not available or very limited, and following an existing standard is a safer choice than deciding on something else with no proof.

The same philosophy was applied when selecting the material which is **nylon materials** like **PA11** or **PA12** to be preferred. Case studies have shown that it is more lightweight and flexible compared to the traditional plastic used as a sheet material and similarly dynamic like carbon fibre used in AFOs [41] and even biodegradable [44], contributing to an environmentally friendlier impact.

Overall, the combination of a SLS printing method and the nylon material results in a positive side effect of **reducing production costs** by up to 50% by saving up in manufacturing time and material waste [41].

When creating the design, the first step is to make a **3D scan** of the patient's legs to gather measurements, which is replacing the traditional process of taking a cast. It is important to keep the ankle at a 90° angle, which can be achieved by placing the foot on a board of clear acrylic like shown below in Fig. 56.

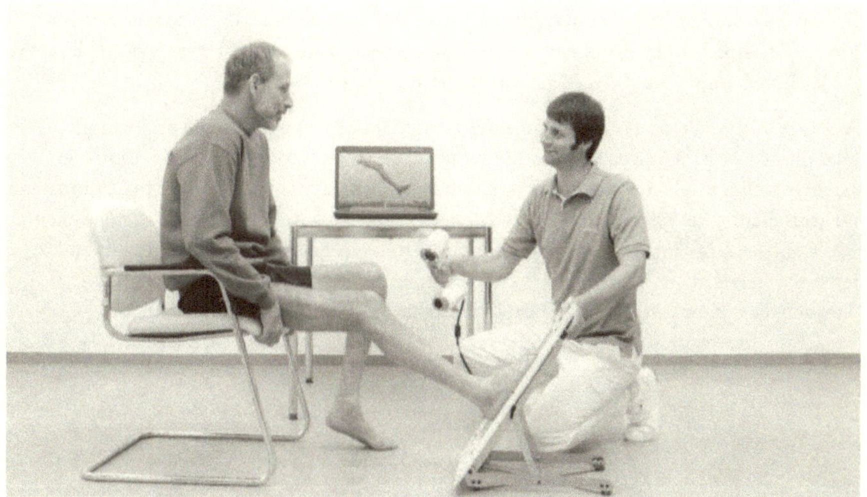

Figure 56, 3D scanning method for taking measurements for an AFO [49]

The next step is to make a CAD model with 3D modelling similar to the process of the PROTOTYPE phase in this project. However, not every orthotist can be expected to excel at 3D modelling or be capable of learning it, moreover, there is no need to make each AFO from scratch. Therefore, a program should make use of algorithms and pre-sets to speed up the process efficiently. A base design is then wrapped around the 3D scan of the leg, further adjustments can be done if the orthotist as an expert wishes to do so, depending on their judgement over the patient's condition that can vary from person to person.

Lastly, a pattern will be translated to the surface using parametric design that uses algorithms defining the maximum size of a cut-out, how a pattern should spread and disperse, and borders of the canvas where it must stay within. If a custom pattern is desired, further assistance of a designer may be required.

Finally, when the 3D model is finished, the patient can choose their desired colour and surface finish from a broad material library. If a rendering engine like Keyshot is in use, it is possible to generate real time previews to confirm the design with the user before ordering, meaning that they have been involved in the design process all the time, guaranteeing high customer satisfaction once the finished ankle foot orthosis arrives.

5.3   Physical Model

The physical prototype produced is a mere appearance model that cannot withstand the load of being worn or twisted at all. The purpose of making a prototype of this kind nonetheless is for its communicative value. Perceiving a product not only visually but also by haptics can change a user perception greatly and adds to the user experience. It is a great tool to present to the client, teachers, professors, and colleagues, expressing how the AirWalker design could be if it was a real commercial product.

Another advantage of having an appearance model in this case, was the opportunity to take photos of it. When cautious, wearing the appearance model for taking photos in context with a model is possible as well. Since the designer is also the user providing the measurements, the photos were taken by a colleague or with a timer while the designer had to pose as the model. It was advantageous to have a model with a perfect fit and to have the freedom and flexibility to schedule a photo shoot whenever it was needed without relying on external help. Some photos representing the intended looks of a working AFO made from the intended printing method and material are shown below in Fig. 57, Fig. 58, and Fig. 59.

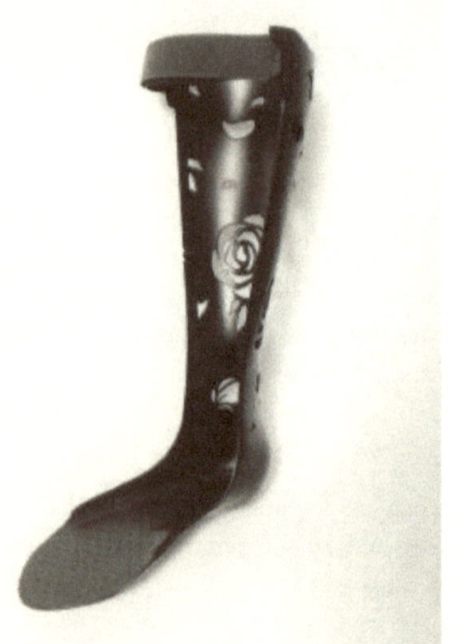

Figure 57, Finished appearance model, front (left) and back (right)

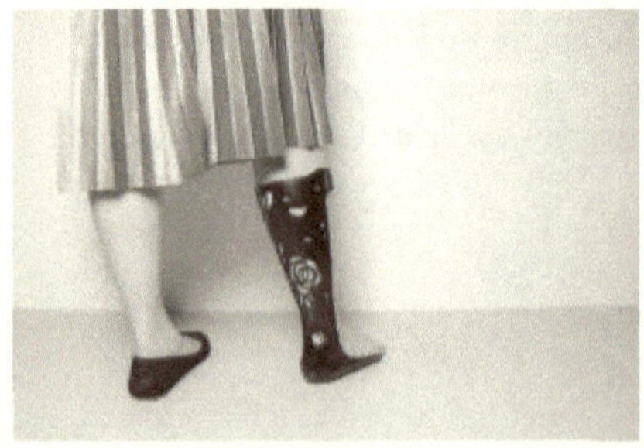
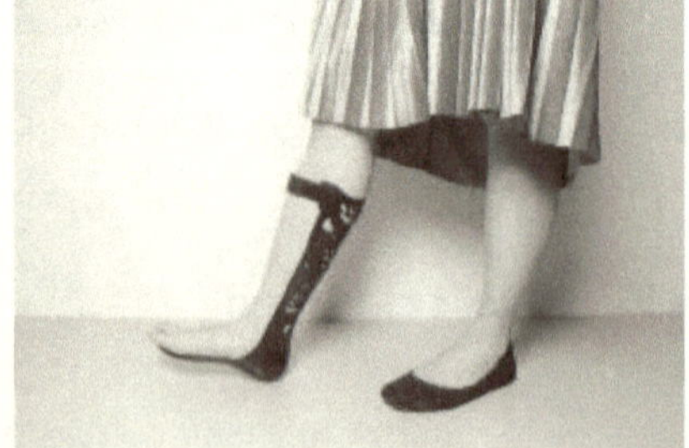

Figure 58, Appearance model worn by the designer, whose leg measurements were the base for the CAD model and prototype

Figure 59, Detail shots of the top part with Velcro (left) and sole perforation (right)

The specific pattern and colour choice for the appearance model was made by the user and designer of this project themselves, therefore, the design choices were influenced by what they thought would fit their style and rest of the wardrobe best. The rose pattern was their personal favourite that represents a feminine fashion theme that reflects in other frequently worn outfits as well. Although it is possible to pick any favourite colour, a more neutral one like black was preferred, so it is easier to match it with daily changing outfits and footwear that is mostly black. Matte as a texture was chosen because a matte surface has better properties at visually hiding minor damage like scratches and bumps which are likely to happen to a product that is worn under loads and exposed to the environment.

5.4   Design Variations

To express the diversity that is possible to implement for AirWalker, design variations that were modelled for a more feminine and romantic theme, a more masculine, neutral, and cool mood, and for a playful theme aimed at young kids are shown below in Fig. 60.

Figure 60, Design variations expressed through patterns for males, females, and children (left to right)

Some details of each theme and how their patterns look on an ankle foot orthosis are shown in close-ups in their respective collages in Fig. 61, Fig. 62, and Fig. 63 which are presented below. The first figure with the feminine rose pattern was also the design used for the prototype and appearance model since it was reflecting the designer – who had to wear it for photos and testing – best, however, the paint colour was changed during prototyping.

Figure 61, Details of the rose pattern design, aimed at a feminine theme

Figure 62, Details of the Kumiko pattern design, aimed at a male theme

Figure 63, Details of a space pattern design, aimed at children

Apart from the shown designs, more variations are possible, which depend on the patient's preferences. For this project, far more patterns have been designed and drawn as vector graphics, but they have not been applied on the 3D model. A selection of the eight most popular motives is shown below in Fig. 64 and the whole catalogue with all 21 patterns can be found in the appendix.

// **CHILDREN** – TOP 4

// **ADULTS** – TOP 4

Figure 64, Selection of most popular patterns, based on a Kansei Study conducted online

6 Conclusion and discussion

6.1 Discussion of Methods

All methods for research and gathering information were very effective for this project. In the current age of digitalization, it is essential to use sources on the internet as well since many companies prefer to publish case studies online and research papers from overseas are not all available in the university's library. Most of the relevant sources and papers were found through the search engine Google when the right keywords were applied.

An efficient amount of time was spent on ideating through sketching both traditionally and digitally, depending on what seemed to be the most convenient and fast method in each situation. Ideation then quickly moved on to use 3D modelling to create a 3D model, which was a major product to finish for this project, for both presentation and prototyping. Having the freedom to pick a software of choice for CAD and using Rhinoceros 3D where experience and expertise was available, saved time and enabled an efficient workflow. Making the 3D model was simple and straightforward; the result was of a high enough quality to print an appearance model with smooth surfaces. As a result, three variations with different patterns could be produced for the final presentation which made a more positive impact than just showing one variation.

Unfortunately, offline testing and user research sessions were impossible to conduct after March due to restrictions in the maximum number of participants in gatherings and rules for social distancing. The Kansei Engineering study that was supposed to be held in a focus group was difficult to conduct when abiding by those rules. Instead, it was moved from a focus group to an online questionnaire, which did not bring the same expected result and was difficult to handle in its own way but delivered the closest possible data to what was desired. Luckily, many important user interviews were taking place before the restrictions were effective, or they have been conducted virtually.

Finally, the methods for making the appearance model and for rapid prototyping could have been more fitting to the intended method of the concept, but since this project had no funding behind it, it was the most cost-efficient way using FDM printers that were available and free to use for projects on campus. Thanks to the many printers available on campus, the failed print attempt of printing in one piece was unfortunate, however, smaller printers were immediately available as a back-up and splitting the model into three pieces was an effective alternative. Perhaps, splitting the model up came with an advantage that was not considered before which is the easier handling of the prints when sanding the surfaces by hand. Holding the print with one hand while sanding with the other would have been difficult with a model thrice the size of each individual piece that resulted from splitting the model.

All in all, most methods used were as efficient and effective as possible until the outbreak of COVID-19 in March 2020 made all offline activities difficult to conduct. Alternatives then implemented were not always the ideal methods, however, they were the best ones available under the given circumstances that were force majeure. Because this project was running on a very small or rather no budget, there was no access to the newest printing techniques, not even through ordering from third parties. In retrospective, that could have added significant value to the prototype and the project result by using the intended production method and material to make the model with. A functional prototype could have also served for a quick case study that would present new leads for future improvement. Because additive manufacturing is all about making one product at a time with low set-up costs anyway, it would be advisable for any future projects in this field to try to receive funding, so the right printing technique can be tested and other tests can be conducted as well.

6.2   Discussion of Results

Before the project has started, the client set-up three research questions that are the main research topics. The purpose of this project was to re-design an ankle foot orthosis in a way that utilizes additive manufacturing for production, to explore new possibilities and advantages over the traditional product and its manufacturing process. To discuss the results, they need to be evaluated upon how well they answered those research questions.

To answer the first question '***How can an ankle-foot orthosis be made aesthetically pleasing whilst still meeting functional needs?***', the concept for AirWalker shows that the change in the production method from all traditional techniques to additive manufacturing is the key that opens up new possibilities for aesthetic and functional improvement.

The old AFO was made in the most efficient way to cut up sheet material, it was built with the idea of splitting and taking off material. If 3D printing is 'additive' manufacturing, then the traditional method could be described as 'subtractive'. The additive approach of manufacturing nullifies all previous limitations in geometry and enables the design to integrate all design elements into a more harmonic form, thus meeting all the functional needs in a more aesthetic way. Since the framework of the old AFO has been mainly kept and personalization options are implemented within that 'frame', the past functionality remains while new functionality like flexibility or air-permeability and aesthetics enhance the product value additionally.

Unfortunately, the technical feasibility cannot be guaranteed as it was not tested and a specific plan to carry out the implementation for production does not exist since the time for this project was limited. However, according to other reports and direct competitors, 3D printing is already used successfully in orthotics and that is evidence enough to validate the feasibility of this conceptual design.

The same key that is additive manufacturing also answers the next question which was '***Is it possible to make them look more aesthetical or personalized so that they reflect each patient's personality?***'. The answer is 'yes' and again, thanks to 3D printing it is possible to change up the geometry in a simple way that was not possible before. The approach that is building layer upon layer makes it possible to implement cut-out patterns on the surface of the AFO that would have been too tedious and not worth it if an AFO was made in traditional ways, that was a mix of buying as many mass-produced items from a factory as possible and making the rest of the AFO completely by hand.

The currently existing personalization method involving the transfer papers or foils rely on what mass production of them has to offer. Therefore, the range is limited to the specific catalogue of illustrations that a doctor has access to, patterns often focus only on children's preferences while the few options for adults are rarely appealing. Furthermore, the warping that often happens around the ankle, distorts the pattern to a point where it is not pleasant to look at anymore. The final concept of this project allows personalization expressed in patterns that are cut-outs, 3D modelling gives more control over the pattern flow, any customization is more feasible and affordable since it utilizes curves only or black and white shapes and not full colour illustrations anymore, and the patterns are not overly colourful and distractive so that adults can enjoy a subtle and elegant element of self-expression.

To go into more detail of the self-expression opportunities and also to answer the last research question which was *'What can ADM as a manufacturing technique offer orthotists and patients in terms of choices and self-expression?'*, it can be said that thanks to the ease of producing the physical product through 3D printing, more time and energy can be invested on designing the ankle foot orthosis to the taste of each patient. The concept for AirWalker puts emphasis on customization and the production method makes one product at a time only, so it encourages to explore all options of customization and self-expression to make each individual patient's unique ankle foot orthosis.

With the push of one button only, the 3D printer will translate a digital file into a physical product that only needs little to no post processing and constant surveillance over the printing procedure is not necessary. The manufacturing of the actual product will run as easy as never before, but for that it is crucial to have a file free of errors to avoid reprints to save time and material.

To create that error free file will be a new responsibility of an orthotist. The extra time gained from eliminating traditional steps like making the casts can instead be invested in designing the best fitting AFO for each patient, in both a physical and a stylistic sense. Furthermore, a patient can be invited to involve themselves, becoming the designer of their own device and therefore having more influence on the outcome, which is a great progression in terms of choices and self-expression. On top of being able to work together with the patient, an orthotist can now focus more on every individual's rehabilitation plan as well, which was one of the interviewed user's wishes.

Another significantly changed design aspect thanks to the new possibilities that additive manufacturing offers is the now minimalized device thickness. Thanks to that, it is now up to the user to decide whether they want to hide their orthosis and their disability – which was never possible before unless wide pants were worn – or show it off and wear it with pride. More than being able to design the pattern on the AFO, a patient can now decide how much of their disability they want to disclose. According to user interviews and blog articles, not every person's personality is extroverted enough for showing off their AFO and many introverted people exist that make a great effort to hide their orthosis but find that difficult [4]. The idea to hide the AFO as much as possible is similar to the invention of contact lenses, so people with seeing deficiencies can eliminate the prominent looks of eyewear on their face while still being able to see clearly. The existence and popularity of contact lenses proves that hiding a disability is a crucial cosmetic need for users even though so many design variations exist in eyewear, which shows how life-changing it could be for a patient if their orthosis can be hidden.

To achieve all the changes mentioned above however, a great shift from traditional methods to digital methods is needed. To fully implement a 3D-printed ankle foot orthosis as a service and a product, means to rely on digital tools which could not only change the future of the patient but also the future of all orthotists. New equipment and expertise are required that is not what an average orthotic clinic and their employees have nowadays. Tools like computers, 3D Scanners, and CAD programs must be purchased, and a 3D designer or product designer must be hired, unless the future teaching curriculum for orthotist changes and they are taught how to use CAD.

In the end, as promising the presented result may be, when it comes to testing its validity, the evidence is lacking since the results are very conceptual. The solely aesthetically functional appearance model could not have been used for any functional testing and due to time limits and legislative regulations in the Spring semester of 2020, it was not possible to conduct another Kansei study or focus groups with AFO users either. Those studies would have given the concept a more realistic and in-depth approach that eventually could have stirred the outcome into another direction.

All in all, the professor & teachers, the client, and the designer themselves judge that the new concept indeed moves to a direction that looks more aesthetical compared to the traditional designs, and a significant improvement has been achieved with the result. The results are satisfactory and of a good quality for the time spent on it. The probably only but also the biggest loss of this project is that the feasibility and functionality cannot be proven. However, that was to be expected from the beginning that such a test study would be impossible to conduct without further help, funding, collaboration, and on top of all: time. Therefore, the result presented as it is, is satisfactory because it has answered and elaborated on all research questions, utilizing all the human centered research and design methods possible to gather relevant data that proves the validity of the concept.

6.3   Recommendations for the future

It must be noted that this project's result is conceptual to the point that it encourages further research but the research done so far has proven the need and importance of a change in the orthotic industry that deserves to be implemented instead of being treated as a conceptual design of a book project only.

One innovative step imaginable for the future of AirWalker is to eliminate the Velcro part as it is now and print a fastening method that is fused together with the AFO shell so the whole product can be printed in one go and does not need an additional step in the production line for the strap to be completed. An idea is to use the geometries of the so-called mushroom head Velcro or DualLock Velcro as it is called by the company 3M. Its characteristics are that both parts of the Velcro are the same, unlike the traditional hook and loop type, and they are made from a solid plastic material with dimensions bigger than the usual hooks. The stems on the tape have mushroom shaped heads that interlock with each other while giving an audible feedback when successfully locked together. An example of the Velcro and how the interlocking works are shown below in Fig. 65. Thanks to the scale of the geometries of those stems, printing them might be feasible.

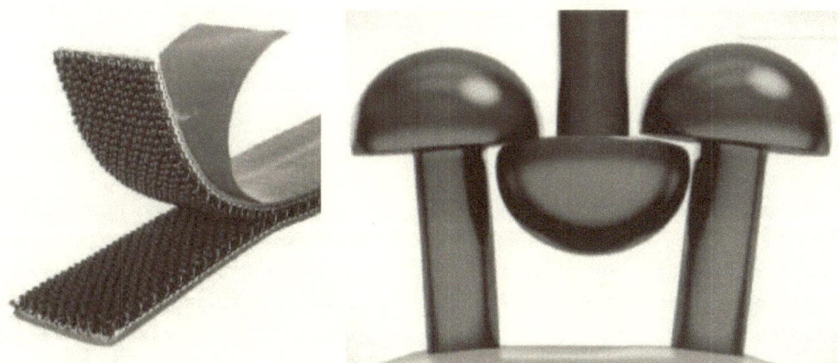

Figure 65, Mushroom head Velcro: a realistic photo (left) and a close-up of the locking system (right)

Another thing consider for the future is to carefully select the sizes of cut-outs for patterns and potentially change of patterns into a raster of lines instead of leaving areas as big hole. The client has voiced their concern over sharp corners possibly pinching the skin during motion and therefore hurting the patient which is a valuable critique that needs future consideration to prevent injuries. An idea to eliminate dangerous sharp corners whilst keeping the wanted illustration and effect of having holes was to apply the Gestalt law of grouping to build up a shape out of several smaller cut-outs that could be dots or lines. In Fig. 66 below, the example was drawn using vertical lines to illustrate a shark, without eliminating its intended shape or size.

Figure 66, Example of reducing sharp corners and allowance to pinch the patient's skin with the initial shape (left) and the rasterized one (right)

It is important to remember that it is often only an orthotist and the patient designing the AFO since a technician, an engineer, or a designer is not always able to join them. Therefore, the development of a new, user-friendly CAD software specialized in making AFOs using automated pre-sets or algorithms is highly recommended, so users with no prior knowledge to 3D modelling can handle it. The lack of an allrounder solution for 3D modelling software has always been a problem, even among designers and engineers, but because every product can require other tools, it is almost impossible to build such a solution. Thus, creating another CAD program for the orthotic field, perhaps even one that is specially catered to AFOs, is likely to be the better result but that involves programmers and software developers, therefore costs, which need to be considered when pursuing a feasible implementation with AirWalker's philosophy, where the patient is the designer as well.

All in all, to make AirWalker a real product, many iterations of designing and testing involving more than just one industrial designer and taking longer than four months are needed. On top of the aforementioned points, it is recommended to look into SLS printing, the actual printer intended to be used, preferably into Multi Jet 3D printers by HP as well, and to conduct more material and user testing since the feasibility of ADM technology for orthotics is still questionable as of now, and printed products will have difficulties keeping up with the stability, endurance, and functionality standards of other factory-made orthoses and finding the right ADM method will need intense testing which was not part of this project. In the future, 3D printing will evolve, and it will bring new possibilities again that might shake up the very foundation of the presented AirWalker concept.

6.4   Conclusion

For the ankle foot orthosis that is a product without any significant innovation in its basic design for 50 years [3, p. 162], this project has uncovered unexplored potential to become a more aesthetically valuable object to the patient, making it an accessory that expresses personality and style despite being a health device, similar to what eyewear is nowadays. The patient gains full control to decide whether to show their AFO off or to hide it and is involved in the design process as much as possible. Aesthetically, the new AFO concept offers endless possibilities for patterns and even textures. It is possible to treat orthotics with a more fashionable twist, designs can change with the seasons and follow trends just like clothes or eyewear, which is just another health device where detailed customization has been made possible for ages.

A positive effect on the economical side is that 3D printed AFOs result in the elimination of tooling costs (time, labour, and material) and quick adaption, meaning a fast turnaround time. Considering that 3D printing becomes more accessible and affordable with every year – especially for the FDM method –, it is to be expected that even the intended novel Jet Fusion technology will become a feasible option for printing orthoses in the future. If a printed AFO becomes affordable enough, a user can even own several pairs and vary them as they want to match their different occasions in life just as they would do with footwear, making it possible to treat an AFO just as any other fashion item. Furthermore, orthoses in general will be better accessible and affordable in lower income countries, enabling equity and humanity around the world by raising the life quality.

Shifting from the traditional manufacturing method to ADM means that two casting steps can be skipped, and a notable amount of plaster material and its waste is saved. The casts that usually have no purpose outside of being used as a mould do not need to be stored or thrown away anymore as there is no need to make them in the first place, resulting in less environmental impact and less material waste with each AFO that is produced. The introduction of materials like PA11 that is biocompatible and better recyclable than materials like carbon fibre adds to the environmental friendliness of AirWalker's design.

In conclusion, AirWalker does not only offer significant enhancement in cosmesis and improved emotional value for the user but also improves the functionality while it allows past advantages to remain. Along with aesthetic and emotional gains, there are economic, environmental-friendly, and technical benefits that pushes the boundaries of technology and science towards experimentation and ultimately innovation. The implementation of digital tools including 3D scanning, CAD modelling, and 3D printing technologies introduces a new kind of convenience and user experience for ordering, adjusting, and replacing ankle foot orthoses. If the process can be simplified thanks to software, it is possible to offer printed custom AFOs that the orthotist and patients can take care of themselves. Having a choice in dressing and appearing in any desired way is a part of self-determination that is a human right many patients with orthoses have lost through social stigmatization, but that right is now being returned to them. Giving a patient the power to eliminate their stigma and turn around their life despite their disability is of an emotionally unmeasurable value. Wearing a physically and emotionally perfectly fitted ankle foot orthosis is not only compensating for their physical shortcomings but will also improve their mental health and life quality forever.

6.5   Project Reflection

Conducting this book project to work and finalize a concept for AirWalker meant working in the field of Orthotics which was an interesting and novel field to investigate, since I have not had any touchpoints with it in my academic career yet. Being a design engineer who knows more about both manufacturing and materials and aesthetics than actual practitioners in the fields of Orthotics, it was an enriching experience for both the student and the client to interchange knowledge and have a conversation about possibilities to bring forward change.

Thanks to the status of being a design student, solutions to this problem were wider explored than if it was conducted as a design project in a company. A variety of techniques and possibilities were considered, however, the resources and funds for execution and testing would have been higher if it was backed up by a company or a client but then the freedom in exploration could have been limited or be under a different kind of pressure. With this being a book project, I was able to take advantage of and insist on using technologies and material potential that practitioners might not have considered yet, while using all the resources the university offers and implementing all the tools that I am comfortable with. Furthermore, I gained more knowledge about my working style and preferences in tools and methods, which will prove to be helpful in my future career.

When reflecting upon the course of this project, it must be noted that many things did not go as initially planned in a large scale. Design projects in general tend to evolve and change as they progress, and flexibility is a must-have for any industrial designer, so ending up with something different than intended is nothing surprising. However, it were major workspace changes and sudden makeshift methods from even before Mid-Presentation that have made this project become so challenging, emotionally and physically, as nobody could have anticipated that a pandemic outbreak in the middle of the academic semester would happen.

A sudden shift in the working environment from the studio to the bedroom and cancelled methods to carry out research and user testing due to social distancing and closed facilities, were the most challenging obstacles. Focus groups were cancelled and visits to the studio became less frequent, leaving me with a lack of information and feedback, while I needed to be creative with how to acquire data and ideate from home. On top of all, keeping up with a large amount of news – globally, socially, and programme-related – while staying safe was additionally mentally exhausting. Processing concerning news added up to stress, pressure, and anxiety. That unwanted distractive nature made it extra challenging to adapt quickly, so I could stay focused and on track with the time schedule.

When it was time to produce the prototype, limited availability of manufacturing methods on campus in terms of ADM were not ideal but workable. Printing with PLA material in an FDM method meant that smoothening and sanding all edges for the model was not a very difficult but extremely tedious task. It would have been more interesting if the prototype had been printed using an SLA or SLS method in one piece instead so the model could have also been functional to a certain degree and acted as a specimen for some testing, putting the value of the appearance model up to discussion. However, there were no funds behind this project and certain printing methods not available on campus. The prototype at hand was nevertheless useful for project photos, especially for shots when it was being worn by me as shown below in Fig. 67.

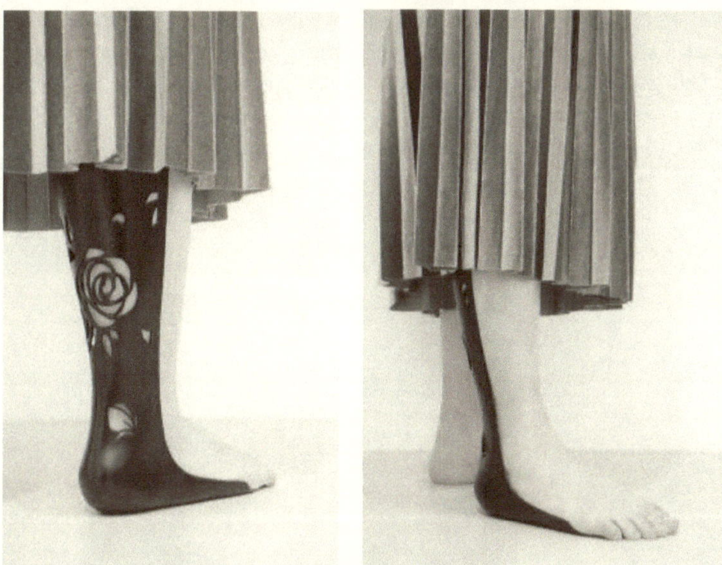

Figure 67, photographs with the worn appearance model in proportion with a user

Considering that this project already started four weeks later than the curriculum intended to, and underwent major changes in schedule, content, and methods, the fact that it was carried out and completed in time nevertheless, shows that I have succeeded in time and resources management by adapting quickly and gained that as a new skill on top of others acquired through practicing during my two years Master studies. The loss in time was compensated by taking available shortcuts and extending working hours to the weekend. The project was manageable in its workload; however, it was extremely demanding and challenging to stay focused, determined, and organized, so all deadlines could be met.

To touch upon the orthotic field that is rarely in the spotlight of industrial design because it is not as commercial as consumer products and lacks publicity, was worth looking into because despite its low popularity, it has one of the most urgent user needs. Shockingly, Orthotics seems to not be widely explored yet in terms of design, which is a shame, however, this was a personally exciting project that broadened the horizon of what Industrial Design and User Centered Design can be.

It was a perfect final project where I made use of all my expertise and experience gained up to this point as I was allowed to use all the tool that I felt comfortable with. Personally, I reckon that I could not have conducted this project to my satisfaction like this without the experience gained in the two years of studying the Industrial Design master at Jönköping university and without the feedback from my teammates, which where students, and teachers and even friends. I am glad to have contributed to the orthotic field with my expertise and hope it will carry on into the future and push forward a movement towards user driven innovation that encourages self-determination and self-expression.

7 References

[1] e. a. Whiteside S., "Practice analysis of certified practitioners in the disciplines of orthotics and prosthetics," American Board for Certification in Orthotics and Prosthetics, Alexandria, Virginia, 2007.

[2] M. M. Lusardi, K. K. Chui, M. Jorge and S.-C. Yen, Orthotics and Prosthetics in Rehabilitation (Fourth Edition), Elsevier, 2020.

[3] M. W. Whittle, Whittle's Gait Analysis, Fifth Edition, Elsevier Ltd., 2012.

[4] S. Pal, "Finding fashion options that accommodate AFOs," August 2013. [Online]. [Accessed 16 April 2020].

[5] D. Norman, The Design of Everyday Things - Revised and Expanded Edition, New York: Basic Books, 2013.

[6] K. Friedman, "Creating design knowledge: from research into practice," Loughborough University, Loughborough, 2000.

[7] A. Milton and P. Rodgers, Research Methods for Product Designers, London: Laurance King Publishing Ltd, 2013.

[8] IDEO.org, "The Field Guide to Human-Centered Design," IDEO, San Francisco, 2015.

[9] T. I. D. Foundation, "https://www.interaction-design.org," March 2019. [Online]. Available: https://www.interaction-design.org/literature/topics/user-centered-design. [Accessed 9 March 2019].

[10] H. G. Fast Company, "Dieter Rams: If I Could Do It Again, "I Would Not Want To Be A Designer"," 2015. [Online]. Available: https://www.fastcompany.com/3043815/dieter-rams-if-i-could-do-it-again-i-would-not-want-to-be-a-designer. [Accessed 2 March 2019].

[11] IDEO.org, "designthinking.ideo.com," 2019. [Online]. [Accessed 9 March 2019].

[12] T. D.school Stanford: Both, "Bootcamp Bootleg," 2018. [Online]. Available: https://dschool.stanford.edu/s/METHODCARDS-v3-slim.pdf. [Accessed 28 February 2019].

[13] B. Rosen, L. Eriksson and B. M., "Affective Surface Engineering – the art of creating emotional response from surfaces," in *15th International Conference on Metrology and Properties of Engineering Surfaces, Charlotte, March 2-5, 2015*, Charlotte, 2015.

[14] ThisIsEngineering, "Person Putting Ankle Foot Orthosis To The Patient," 11 March 2020. [Online]. Available: https://www.pexels.com/photo/person-putting-ankle-foot-orthosis-to-the-patient-3913020/. [Accessed 27 March 2020].

[15] ThisIsEngineering, "Person Wearing Ankle Foot Orthosis," 11 March 2020. [Online]. Available: https://www.pexels.com/photo/person-wearing-ankle-foot-orthosis-3912370/. [Accessed 20 May 2020].

[16] AliMed, "CVA Ankle Brace | AFO Type 670," 2020. [Online]. Available: https://www.alimed.com/type-670-general-purpose-cva-training-brace.html. [Accessed 27 March 2020].

[17] Arizona AFO, "Thermoplastic," 2020. [Online]. Available: https://www.arizonaafo.com/products/thermoplastic/thermoplastic.html. [Accessed 27 March 2020].

[18] AliMed, "AFO | AliMed Carbon Fiber Posterior Lateral Struts," AliMed, 2020. [Online]. Available: https://www.alimed.com/alimed-carbon-fiber-afos.html. [Accessed 27 March 2020].

[19] P. O. Gray and D. F. Bjorklund, Psychology (Seventh Edition), New York: Worth Publishers, 2014.

[20] Interaction Design Foundation, "Gestalt Principles," 2019. [Online]. Available: https://www.interaction-design.org/literature/topics/gestalt-principles. [Accessed 20 May 2020].

[21] R. Monö, Design for product understanding: The aesthetics of design from a semiotic approach, Trelleborg: Skogs Boktryckeri AB, 1997.

[22] I. E. Association, "Definition and Domains of Ergonomics," 2019. [Online]. Available: https://www.iea.cc/whats/index.html. [Accessed 5 May 2019].

[23] General Electric, "What is Additive Manufacturing?," 2020. [Online]. Available: https://www.ge.com/additive/additive-manufacturing. [Accessed 23 May 2020].

[24] Shutterstock, "3D Printer Makes | Shutterstock," 2020. [Online]. Available: https://www.shutterstock.com/nb/video/clip-20702755-3d-printer-makes-printing-bolt-model-screw. [Accessed 29 May 2020].

[25] Amazing AM, LLC, "AM Basics," Amazing AM, LLC, 2020. [Online]. Available: https://additivemanufacturing.com/basics/. [Accessed 23 May 2020].

[26] D. Ullman, The Mechanical Design Process (Fourth Edition), New York: The McGraw-Hill, 2010.

[27] LOFT, "Expert Interview," 2019. [Online]. Available:
https://loft.io/guide/dfa/reference/immerse/secondary-research/expert-
interview/. [Accessed 6 April 2019].

[28] M. Corriero, "The Use of Silhouettes in Concept Design," 2011. [Online]. Available:
http://characterdesignnotes.blogspot.com/2011/03/use-of-silhouettes-in-concept-
design.html. [Accessed 6 April 2019].

[29] F. W. evenant, "Creating Character Concept Art Thumbnails," 2019. [Online].
Available: https://www.evenant.com/design/creating-character-concept-art-
thumbnails/ . [Accessed 6 April 2019].

[30] S. T. W. Schütte, J. Eklund, J. R. C. Axelsson and M. Nagamachi, "Concepts,
methods and tools in Kansei engineering, Theoretical Issues in Ergonomics
Science," Taylor & Francis Group, 2004.

[31] ThisIsEngineering, "pexels.com," 11 March 2020. [Online]. Available:
https://www.pexels.com/photo/making-foot-orthosis-3913022/. [Accessed 29 May
2020].

[32] J. Strömberg, Interviewee, *Expert Interview with a Digital Manufacturing Manager.*
[Interview]. 9 March 2019.

[33] Ottobock, "Digitalisierung und Innovationen für mehr Lebensqualität," Ottobock,
2020. [Online]. Available:
https://www.ottobock.com/de/unternehmen/innovationen/. [Accessed 16 March
2020].

[34] Pohlig GmbH, "Pohlig Printorthesen® – 3D-Drucktechnologie in der Orthopädie-
Technik," Pohlig GmbH, 2020. [Online]. Available:
https://www.pohlig.net/printorthesen/. [Accessed 10 March 2020].

[35] E. Wojciechowski, A. Chang and D. e. a. Balassone, Feasibility of designing,
manufacturing and delivering 3D printed ankle-foot orthoses: a systematic review,
2019.

[36] ot4, "ot4 Orthopädietechnik," 2020. [Online]. Available: https://www.ot4-
orthopaedietechnik.com/. [Accessed 17 March 2020].

[37] M. S. C. Kienzle, "Integration of Additive Manufacturing Processes (3D Printing) in
Orthopaedic Technology Fitting Routine," *ORTHOPÄDIE TECHNIK 05/18,* pp.
48-57, May 2018.

[38] UNYQ, "FAQ's," UNYQ, 2020. [Online]. Available:
http://unyq.com/en/prosthetic-covers/faqs/. [Accessed 20 February 2020].

[39] Crispin Orthotics, "3D Printed Orthotics - The future, brought to you by Crispin Orthotics," 2020. [Online]. Available: https://www.crispinorthotics.com/3d-printed-orthotics/. [Accessed 10 March 2020].

[40] HP, "HP Multi Jet 3D Printing Technology," HP, 2020. [Online]. Available: https://www8.hp.com/us/en/printers/3d-printers/products/multi-jet-technology.html. [Accessed 14 March 2020].

[41] HP, "HP MULTI JET FUSION LEADS TO LOWER COSTS IN ORTHOTICS PRODUCTION," HP , 2018. [Online]. Available: https://enable.hp.com/us-en-3dprint-crispin. [Accessed 20 March 2020].

[42] HP Development Company, L.P., "Transforming prosthetics and orthotics production with HP Multi Jet Fusion technology," 2019.

[43] Scientifeet, "3D Printing," 2020. [Online]. Available: https://www.scientifeet.com/en/home/documentation-en/3dprint/. [Accessed 21 March 2020].

[44] H. Watkin, "Shapeways and EOS Venture into Orthosis and Prosthesis Market," All3DP, 28 March 2019. [Online]. Available: https://all3dp.com/4/shapeways-eos-venture-orthosis-prosthesis-market/. [Accessed 21 March 2020].

[45] TeamOlmed, "TeamOlmed - Ortopedteknik, ortos, protes, ortopediska skor, ont i foten," 2020. [Online]. Available: https://www.teamolmed.se/. [Accessed 4 March 2020].

[46] Mercuris GmbH, "Maßgeschneiderte 3D-gedruckte Prothesen Cover," 2019. [Online]. Available: https://www.mecuris.com/prothesen-cover-3d-druck. [Accessed 10 March 2020].

[47] e. a. Schwarze M., "Wearing Time of Ankle-Foot Orthoses with Modular Shank Supply in Cerebral Palsy: A Descriptive Analysis in a Clinically Prospective Approach," 2019.

[48] UNYQ, "Designs EU," 2020. [Online]. Available: http://unyq.com/en/prosthetic-covers/designs-eu/. [Accessed 12 February 2020].

[49] ottobock.de, "iFab - Ihr innovativer und starker Partner | Ottobock DE," [Online]. Available: https://www.ottobock.de/fachhaendler-techniker/ifab/. [Accessed 26 April 2020].

7.1 Table of Figures

8 Appendices

Appendix 1 Mood Boards

Appendix 2 Personas & Scenarios

Appendix 3 Perceptual Map of Market Analysis

Appendix 4 Kansei Study Questionnaire

Appendix 5 Kansei Study Results

Appendix 6 Pattern Catalogue

8.1 Appendix 1 – Mood Boards

8.1.1 Current AFOs

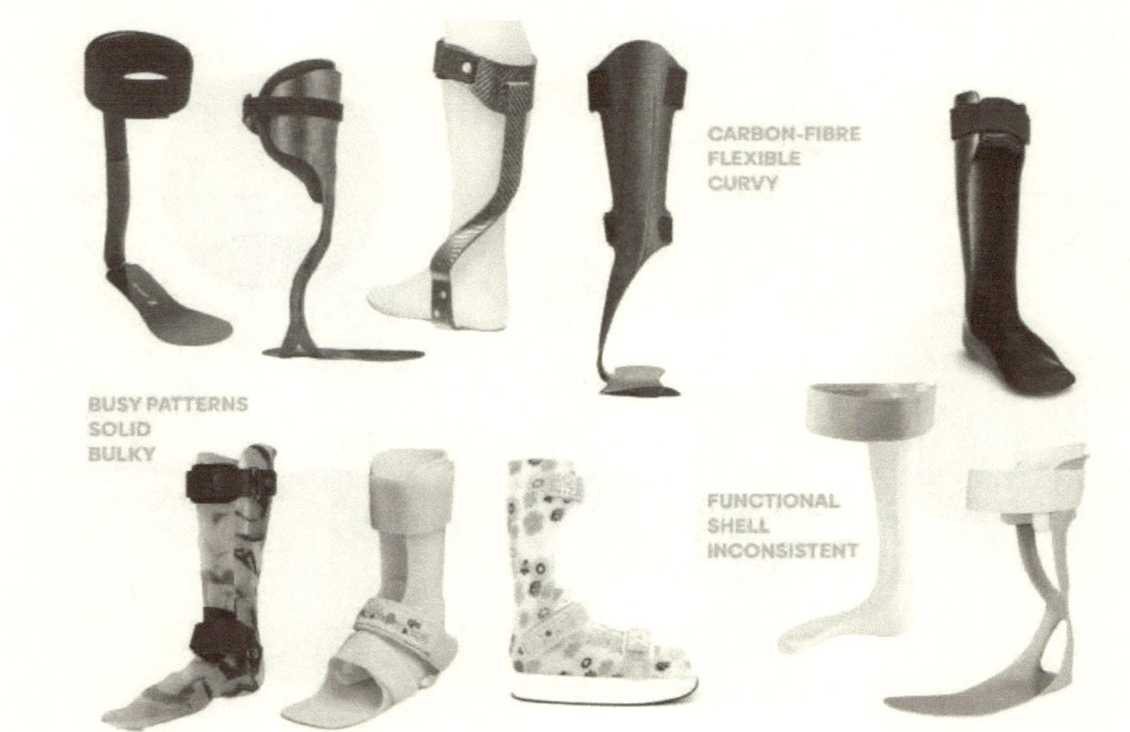

8.1.2   3D printed competitor products

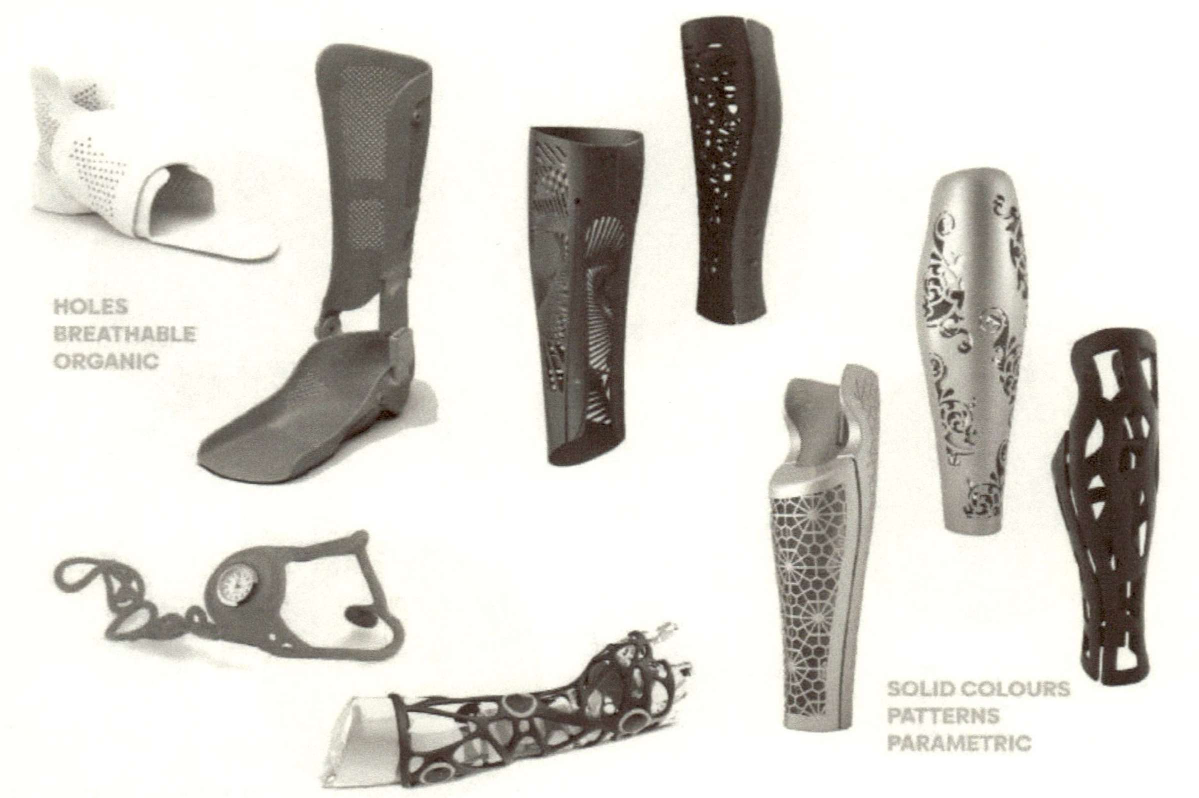

8.1.3 Parametric Design

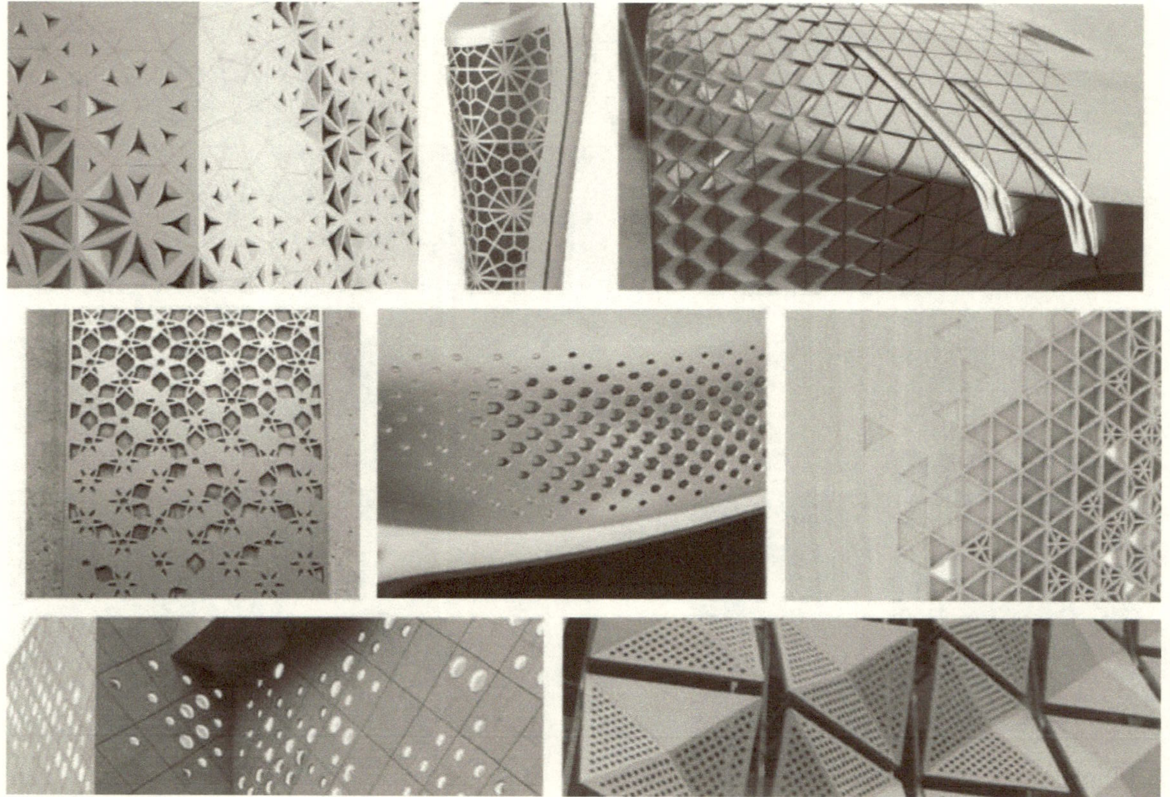

8.1.4 Textures and Straps

8.1.5 Design Language: Organic Curves and Shapes

No distractions
Honest
Minimalistic
'Dieter Rahms/Braun'
Subtle / easy to hide

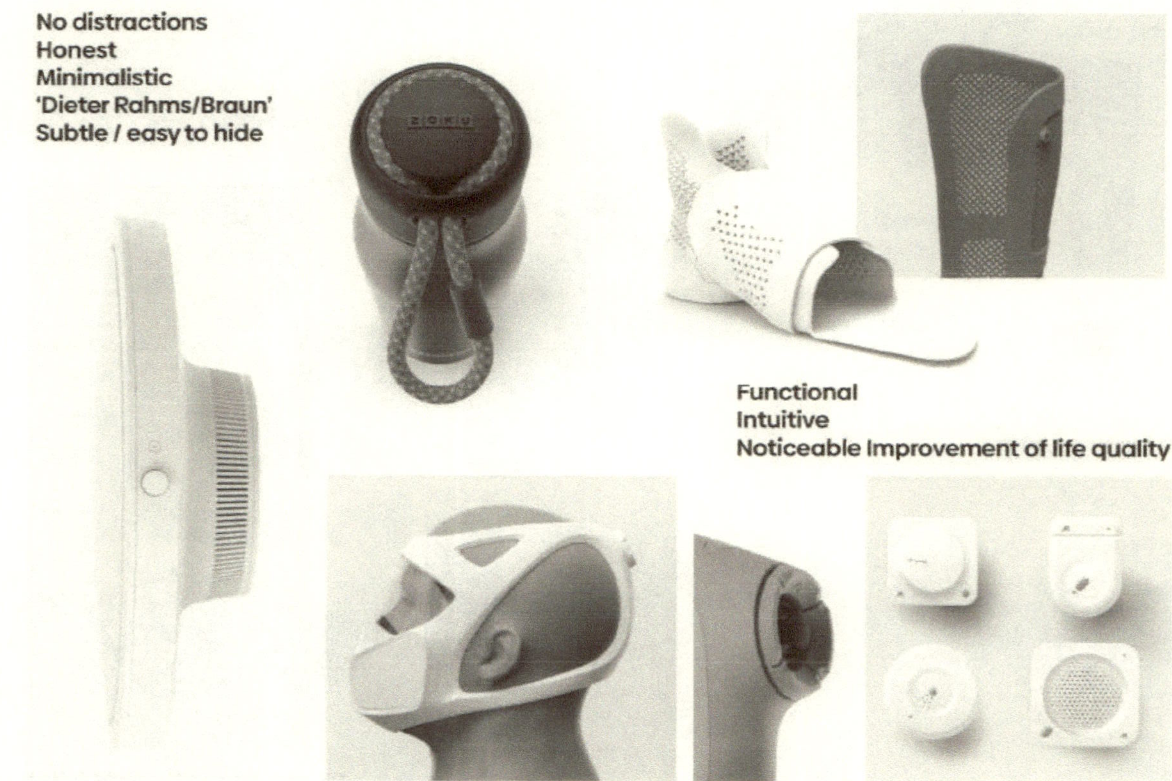

Functional
Intuitive
Noticeable Improvement of life quality

8.1.6 Fashion Mood Board – Children

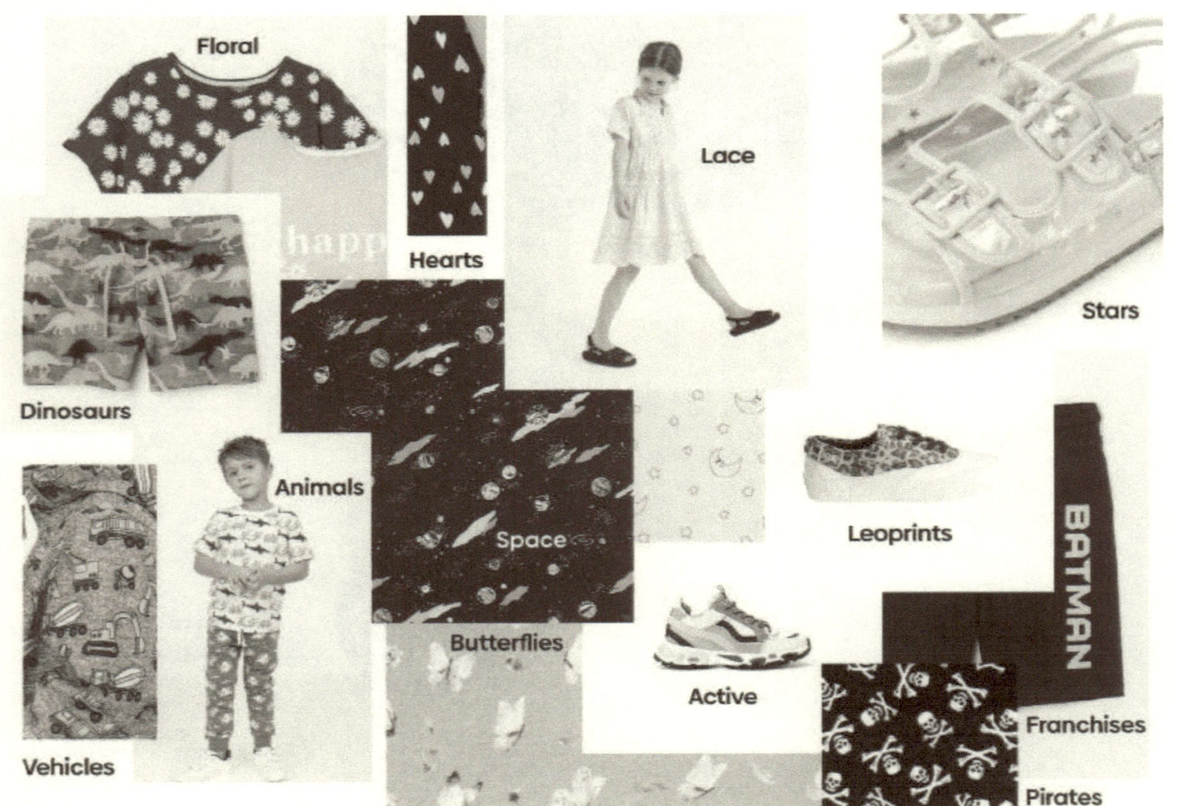

8.1.7 Fashion Mood Board – Adults

8.2 Appendix 2 - Personas & Scenarios

8.2.1 Persona I

LUCAS
8 // The Netherlands

Primary School Pupil

>>> was born with weak legs; has worn AFOs since he was a baby
>>> lives with his parents and older brother and older sister

LIKES

:: football
:: watching TV shows
:: fast food
:: family trips
:: video games with his siblings

FRUSTRATIONS

:: he cannot run and jump as much as his friends
:: starts to feel like he's left out
:: Finds his AFO uncomfortable/does not like putting it on

SCENARIO >>>

Lucas loves sports class in school because it alloes him to move around a lot and he feels like he can play with his friends. However, his AFO is quite stiff and does not allow him to join every excercise with his friends. He tries his best but as a result, the AFO breaks often from him overdoing it. Apart from breaking his AFO, he also needs to get a new fit every year or less because he grows fast but he does not like the procedure at the orthotist or waiting for 2 weeks to get a replacement. Waiting and being at a doctor frustrates him, he would rather like to play in that time like other kids would.

8.2.2 Persona II

PRIYAH
27 // India

Housewife

>>> Has had an accident
>>> Married and lives toge-
ther with husband, children,
and parents

LIKES

:: cooking for herself and her
family, especially likes making desserts
:: playing with her kids
:: meeting friends for tea
:: growing plants in her gar-
den

FRUSTRATIONS

:: cannot handle warm weather
:: her kids start to grow up
and become rebellious
:: doing all the house chores
with limited mobility is tiring
her out

SCENARIO >>>

India's weather is hot and stuffy. Priyah often goes out for a walk
or to do grocery shopping. When walking and excercising her
body for a while, she feels warm and starts sweating, especially
around her calve where the padding from her AFO is. It starts to
get sticky and stuffy on her leg that is wearing the AFO and she
finds it highly uncomfortable and wants to get home asap to
take it off and cool off.

8.2.3 Persona III

CHRISTIAN
36 // Denmark

Web Developer

>>> Stroke patient
>>> Lives together with his girlfriend and his dog

LIKES

:: shopping/fashion
:: video games
:: documentaries on netflix
:: taking walks with his dog

FRUSTRATIONS

:: bad at handling new people/is rather an introvert
:: dislikes the looks of his AFO
:: ashamed of the looks when he wears shorts in Summer

SCENARIO >>>

Christian likes to go buying clothes so he visits the mall on the weekend. When picking a new pair of jeans, he had to visit several stores to pick one that he likes that also is fitting well enough to hide his AFO below. Christian prefers that people do not see right away that he is wearing something that most people do not know of. He handles his disability well but he is tired of explaining it to others every time. Because he knows how it was without it, so he misses the time without the need of an AFO.

CRYSTAL
13 // United States

Student

>>> accident as a child
>>> Single child, lives with parents

LIKES

:: rock music
:: fantasy novels
:: sleepovers
:: sneakers

FRUSTRATIONS

:: feels like an outsider
:: has a hard time figuring out her style
:: buying clothes is difficult for her (fitting & style)
:: wants to make more friends but her disability scares some off

SCENARIO >>>

Crystal is a teenager that is still figuring out who she wants to be and what is her style. Since an accident in primary school, she needs to wear an AFO. In school, she tries to hide that she wears an AFO as much as possible, to the point that sometimes does not wear it on days when she feels better. In school, she gets bullied quite often, some kids label her as 'the disabled kid' and make fun of the looks of her AFO.

8.3   Appendix 3 - Perceptual Map of Market Analysis

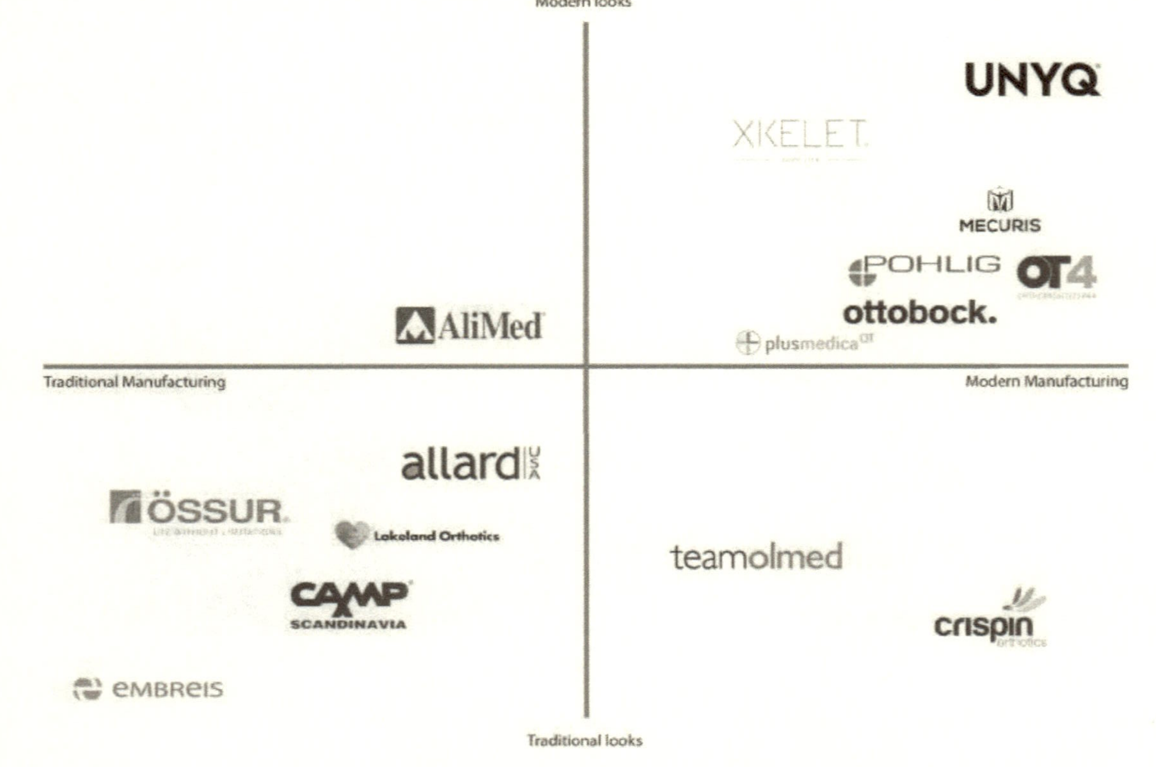

8.4   Appendix 4 - Kansei Study Questionnaire

Page 1

Preferences in Patterns on Wearables

Hello and good day!

My name is Therese and I'm currently conducting my Master thesis for graduating in Industrial Design at Jönköping University in Sweden! The product that I am designing is a wearable health device that helps patients with walking, it is called an 'Ankle Foot Orthosis' Being worn around the lower leg, it has quite a big surface that could be used to give the patient freedom for self expression of their choice.

For my last step I need to validate people's feelings and opinions towards a variety of patterns that could be printed/implemented on the surface of that device. Tastes may differ but all opinions are valid, so there is no such thing as wrong answers, please go ahead and answer with whatever comes to your mind. This test takes approximately 10-15 minutes and its result is solely used to capture first impressions and personal preferences to see if there are tendencies and whether they match up with current trends and expectations.

The questionnaire is completely anonymous and I will not ask for personal information!

Have fun with the patterns and stay safe during these times!

Thank you so much in advance for helping me out!

Next

Page 2 (continued next page)

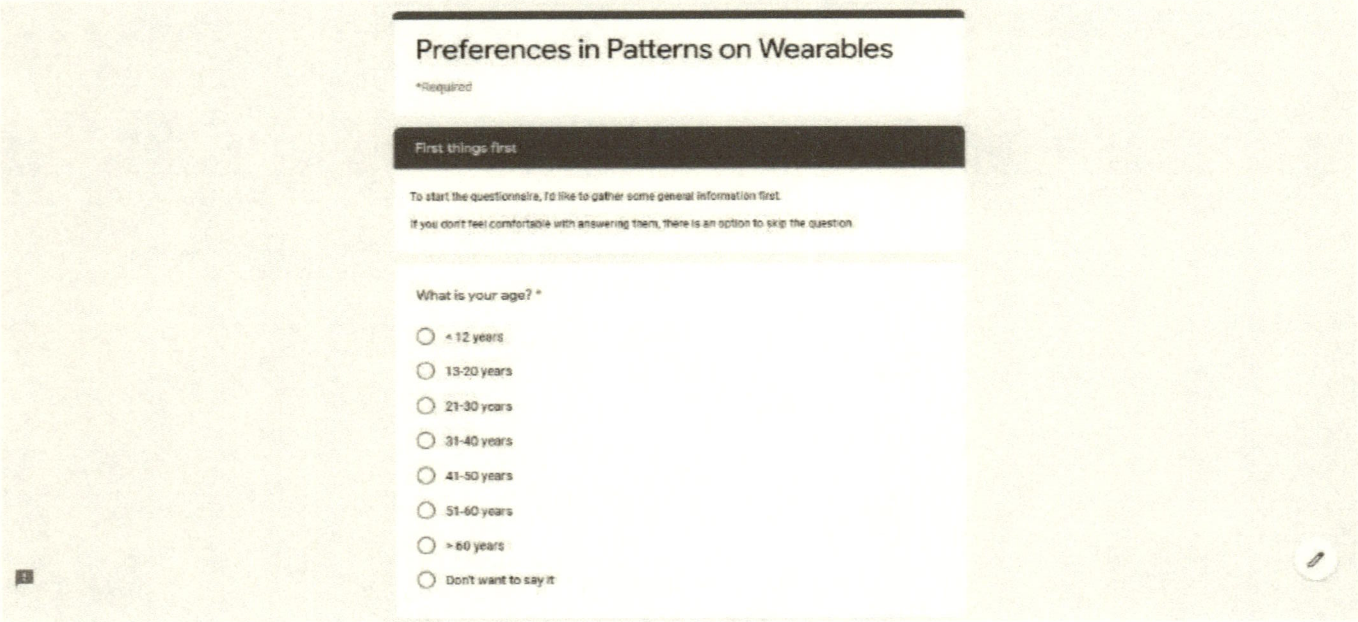

Preferences in Patterns on Wearables

*Required

First things first

To start the questionnaire, I'd like to gather some general information first.

If you don't feel comfortable with answering them, there is an option to skip the question.

What is your age? *

O < 12 years

O 13-20 years

O 21-30 years

O 31-40 years

O 41-50 years

O 51-60 years

O > 60 years

O Don't want to say it

Page 2 (continuation)

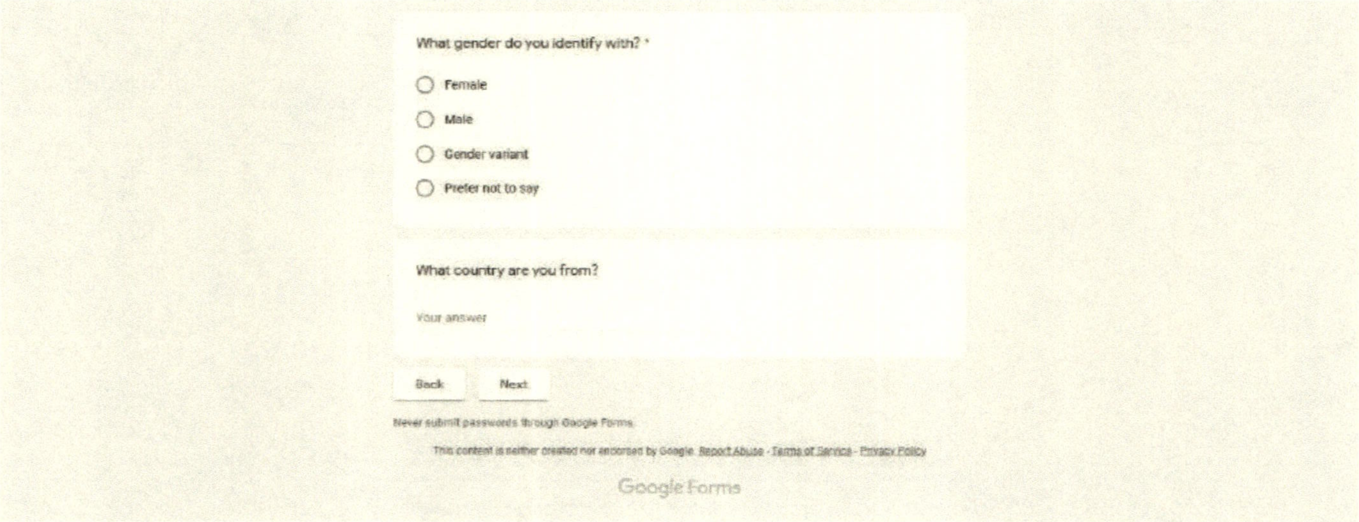
What gender do you identify with? *

- Female
- Male
- Gender variant
- Prefer not to say

What country are you from?

Your answer

Back Next

Never submit passwords through Google Forms.

This content is neither created nor endorsed by Google. Report Abuse - Terms of Service - Privacy Policy

Google Forms

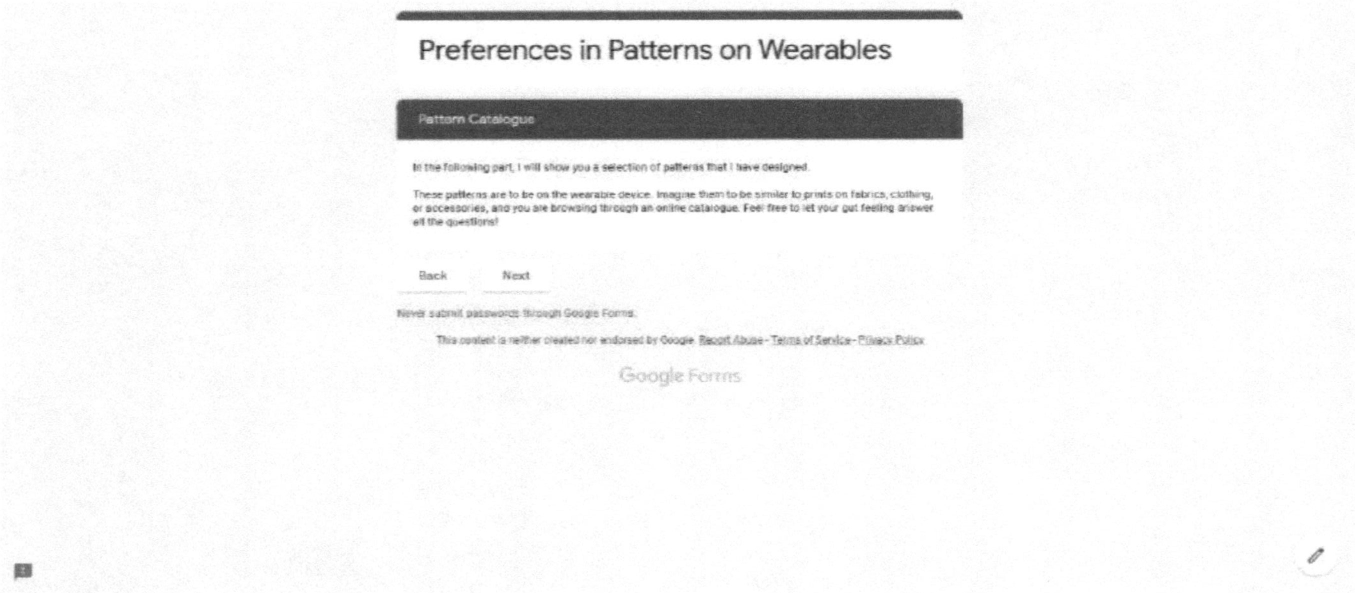

Preferences in Patterns on Wearables

Pattern Catalogue

In the following part, I will show you a selection of patterns that I have designed.

These patterns are to be on the wearable device. Imagine them to be similar to prints on fabrics, clothing, or accessories, and you are browsing through an online catalogue. Feel free to let your gut feeling answer all the questions!

Back   Next

Never submit passwords through Google Forms.

This content is neither created nor endorsed by Google. Report Abuse - Terms of Service - Privacy Policy

Google Forms

Preferences in Patterns on Wearables

*Required

Pattern 01

Pattern 01 out of 21

Page 4 (continuation)

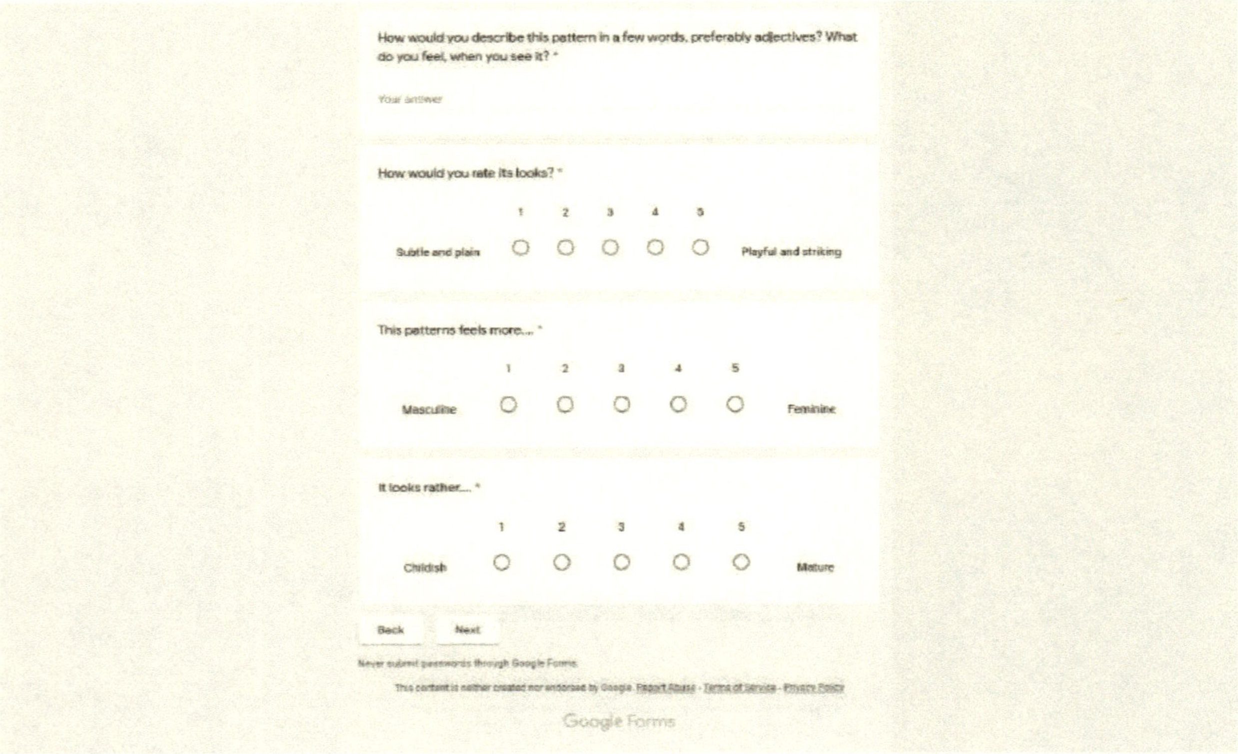
How would you describe this pattern in a few words, preferably adjectives? What do you feel, when you see it? *

Your answer

How would you rate its looks? *

	1	2	3	4	5	
Subtle and plain	○	○	○	○	○	Playful and striking

This patterns feels more... *

	1	2	3	4	5	
Masculine	○	○	○	○	○	Feminine

It looks rather... *

	1	2	3	4	5	
Childish	○	○	○	○	○	Mature

Back Next

Never submit passwords through Google Forms.

This content is neither created nor endorsed by Google. Report Abuse - Terms of Service - Privacy Policy

Google Forms

8.5   Appendix 5 - Kansei Study Results

What is your age?

32 responses

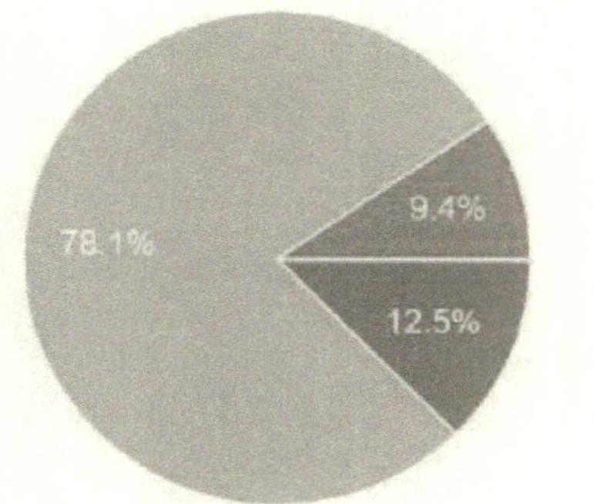

What gender do you identify with?

32 responses

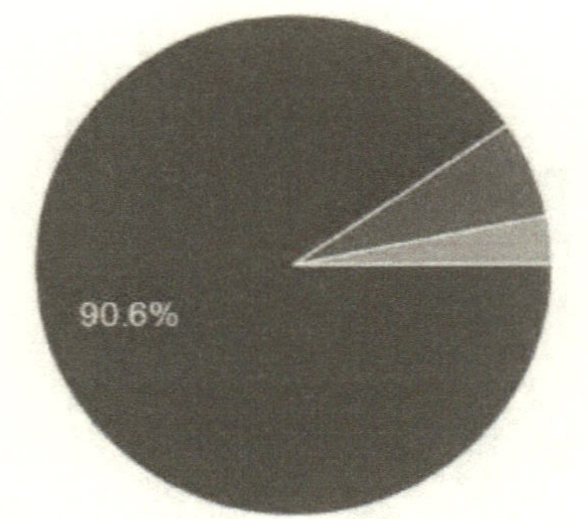

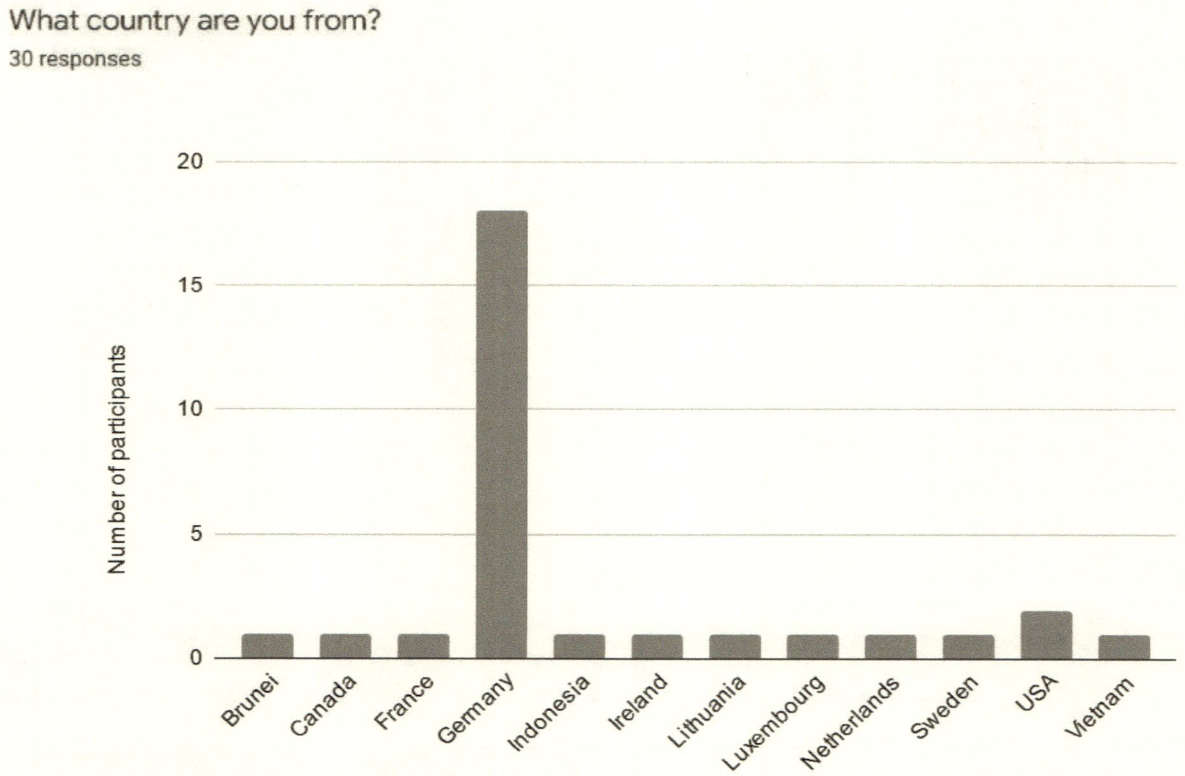

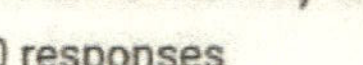

What country are you from?
30 responses
Number of participants
20
15
10
5
0
Brunei
Canada
France
Germany
Indonesia
Ireland
Lithuania
Luxembourg
Netherlands
Sweden
USA
Vietnam

Popularity vote!

Please pick up to 5 designs that you personally liked the most and would wear yourself if it was printed on clothing or accessories (colour variations possible).

32 responses

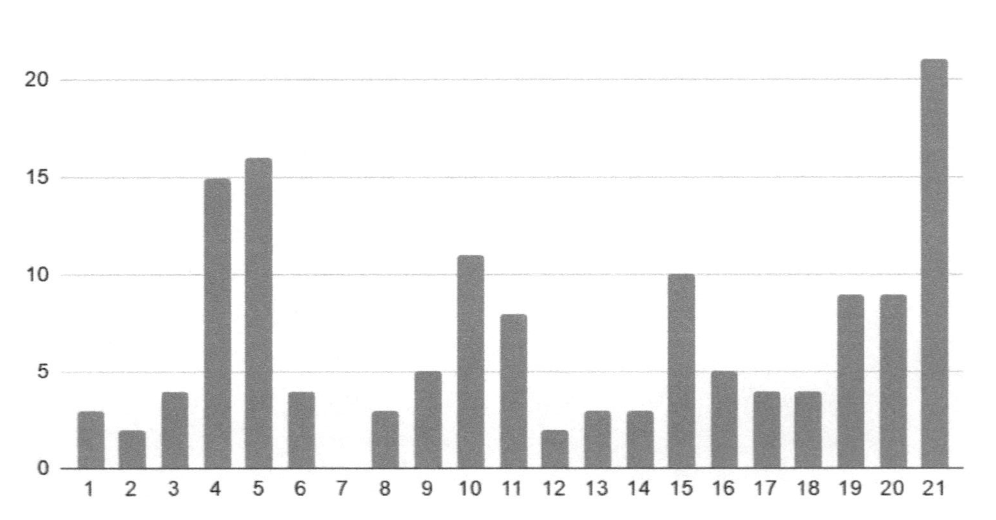

8.6   Appendix 6 - Pattern Catalogue

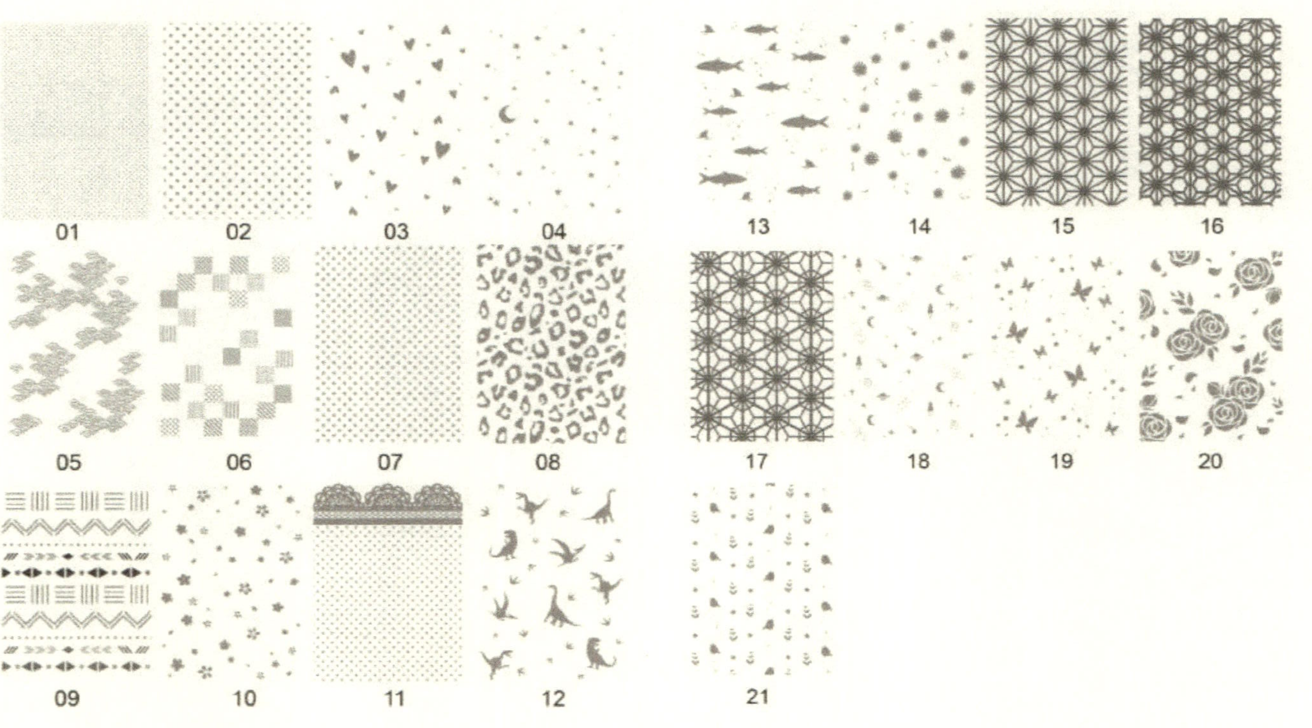

01   02   03   04   13   14   15   16

05   06   07   08   17   18   19   20

09   10   11   12   21